摩登样板间 III
新田园

MODERN SHOW FLAT III
NEW PASTORAL

ID Book 图书工作室 编

华中科技大学出版社
http://www.hustp.com
中国·武汉

前言

浅谈样板间设计

样板间是地产开发商为吸引目标客户而精心打造出来的理想家居空间。设计师的设计重点在于营造引人入胜的视觉效果、注重表现楼盘的特质及展现目标客户理想的生活品位。

在确定设计主题的过程中，我们首先会与房产商沟通，了解楼盘的特质、当地的文化背景以及目标客户群的需求。关注案子本身的售价，以及开发商想要销售的群体。挖掘这一类目标客户群的使用要求以及兴趣爱好，并思考可能会打动他们的生活场景等。

设计流程的先后次序分别是建筑外观、内部结构、空间设计，前两部分由建筑师完成，后一部分则由室内设计师完成。但建筑师并非就具备空间设计的经验，这会导致空间结构布局的不合理，因此在设计之初我们也会介入建筑的设计，使建筑外观与空间结构布局都更加合理。使得我们在之后的室内设计中更加得心应手。

相比而言我们更注重平面，认为平面比立面要重要，因为我们在做平面的时候就把立面考虑到了，这样当我们把立面拉起来后就已经是很好的作品了。我们常使用"灰空间"的手法。利用室内与其外部环境之间的过渡空间，来达到室内、外融和的目的。用"灰空间"来增加空间的层次，协调不同功能的建筑单体，使其完美统一。改变空间的比例，弥补建筑户型设计的不足，丰富室内空间。

总的来看，我们对于样板间的设计有三个原则。扬长，充分展示自己的优点。避短，通过设计的手法来弥补户型的缺憾，房子或多或少都存在某些缺憾，需要通过设计给予弥补或掩饰。主题风格，样板间设计必须有明显的主题思路或风格，让人们记忆深刻。

台湾大易国际设计事业有限公司·邱春瑞设计师事务所
邱春瑞

摩登时代的Modern家居

我对风格的划分一向持保留态度,尤其在室内设计上。倘若将新中式、新简欧及新现代、新田园放在一起,冠以"Modern"一词结集出版,我亦不觉有丝毫不妥。因为它们都有一个"新"字,只要不钻牛角尖,但凡是新的,理所当然都可以认为是Modern的。

除了"新",Modern还应该具有以下的特征:新奇的、时尚的、合时宜的。当年我们看卓别林的《摩登时代》,尽管有一些冷幽默的成分,然而,它毕竟阐述了那个时代的精神语言,而我们在设计概念上注入摩登的成分,目的也是显而易见的,你说我们流于俗套也好,你说我们标新立异也好,最重要的是,我们所结集出版的这一系列作品,毋庸置疑,是这个时代,不,是当下乃至再稍后一段时间国内设计的代言。其中的作品,无论新中式的典雅,新简欧的浪漫,还是新现代的飞扬,新田园的清新,都不约而同地在应和着当今各类业主的需求,也展示着一批会思考的优秀设计师对Modern的理解与把握。

中国古语云"识实务者为俊杰",一个数年致力于设计领域书籍出版的团队,是以极为务实的态度去甄选这套Modern家居系列作品,无论从设计思维还是操作的层面,这套书籍都有可圈可点,可以借鉴、学习之处。

作为设计师,我们通常以上下五千年、纵横千万里、信马由缰、神思飞扬而自诩,而更多时候,也许需要执案自问:我们的思维是否真正Modern了?

香港方黄建筑师事务所

方峻

目 录

006-090

- 006 绿野仙踪
- 012 金地新外滩壹号样板间
- 018 华林上景
- 026 万科双月湾TA2户型
- 032 万科双月湾TB2户型
- 036 星雨华府
- 042 海口美林谷度假村
- 048 让时间放慢脚步
- 058 御湖半山
- 066 芬芳流淌
- 072 桂丹颐景园高层样板间
- 078 鸢尾花开
- 084 城市花园
- 090 游于艺的Feeling

098-154

- 半岛泉水欣座3号 098
- 中冶尚园37幢 102
- 东湖京华京玉苑 108
- 雅居乐·情蓄向日葵 114
- 书香绿苑·钟鼎山庄 120
- 日湖花园某住宅 126
- 半岛时光·城市花园22幢 132
- 世茂外滩新城 138
- 金茂四季花园12幢 142
- 加州阳光 148
- 金海湾某宅 154

160-250

- 160 金陵尚府
- 166 卷珠帘
- 176 水岸风情
- 182 细软时光
- 188 玲珑
- 196 清浅时光·鸿雁名居
- 200 世茂四期5幢
- 206 乡居岁月
- 214 香江枫景
- 220 太子山庄某住宅
- 226 银城西堤国际某宅
- 232 托乐嘉
- 238 左右阳光
- 244 天正湖滨
- 250 中冶虞山尚园

256-316

- 皇冠国际 256
- 中山星汇云锦3栋03户型 262
- 佳兆业鞍山水岸华府GC-A3样板间 268
- 中信红树湾 274
- 星梦奇缘 280
- 蓝湖郡联排别墅 286
- 葱荣岁月·雅致如歌 292
- 太古城D座A户型 298
- 颐慧佳园样板间 304
- 花里林居 310
- 大华南湖公园世家 316

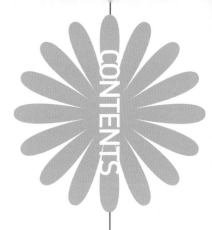

CONTENTS

006-102

- 006 The Wonderful Wizard of Oz
- 012 Gemdale New Bund of No. 1, Show Flat
- 018 Hualin Shangjing
- 026 Vanke Lunas Del Mar, House Type TA2
- 032 Vanke Lunas Del Mar, House Type TB2
- 036 One Residence of Silver Rain Mansion
- 042 Haikou Meilin Valley Holiday Village
- 048 Slowing down the Pace of Time
- 058 Royal Lake Mid-Levels
- 066 Flowing Fragrance
- 072 Guidan Yijingyuan High-Rise, Show Flat
- 078 Blossoming Flower-de-luce
- 084 City Garden
- 090 Feelings Swimming in Art
- 098 Peninsula•Quanshui Xinzuo, Building No. 3
- 102 MCC Shangyuan, Building No. 37

108-176

- East Lake Jinghua Jingyu Garden 108
- Agile Properties•Love of Sunflowers 114
- Scholarly Green Garden•Zhongding Mountain Villa 120
- One Residence of Rihu Garden 126
- Peninsula Time•City Garden, Building No. 22 132
- Shimao the Bund New City 138
- Jinmao Four Season Garden Building 12 142
- Sunshine of California 148
- One Residence of Golden Bay 154
- Jinling Capital Metropolis 160
- Bead Curtains 166
- Waterfront Charms 176

182-256

- 182 Refined and Mild Tim
- 188 Exquisite Residence
- 196 Light and Mild Time•Hongyan Mingju Residence
- 200 Shimao Property, Phase 4, Building 5
- 206 Time in the Village
- 214 Fragrant River Scenery
- 220 One Residence of Taizi Villa
- 226 One Residence of Yincheng Xidi International
- 232 Talege
- 238 Left and Right Sunshine
- 244 Tianzheng Lakeshore
- 250 MCC Yushan Shangyuan Garden
- 256 Crown International

262-316

- Zhongshan Starry Winking, Building No. 3, House Type 03 262
- Kaisa Anshan Waterfront Mansion, GC-A3 Show Flat 268
- CITIC Mangrove Bay 274
- Star in My Heart 280
- Blue Lake County Townhouse 286
- Memorable Years like a Song 292
- Taigu City, Building D, House Type A 298
- Yihui Jiayuan Show Flat 304
- Neighbor in the Flower 310
- Dahua Group South Lake, La Park 316

摩登样板间 III
新田园

绿野仙踪
The Wonderful Wizard of Oz

设计公司：重庆品辰设计
设 计 师：庞一飞、殷正毅、代曼淇
项目地点：重庆
项目面积：92 ㎡
主要材料：环保肌理漆、花岗石、实木百叶、仿古砖、金花米黄大理石、马赛克等

Design Company: Chongqing Pinchen Design
Designers: Pang Yifei, Yin Zhengyi, Dai Manqi
Project Location: Chongqing
Project Area: 92 m²
Major Materials: Environmental Paint, Granite, Solid Wood Shutter, Archaized Brick, Perlato Svevo, Mosaic Tile

耳熟能详的谚语："East, west, home's best."（金窝，银窝，不如自己的狗窝）出自弗兰克·鲍姆的《绿野仙踪》。在这套作品的设计中，流露出对温暖、质感与童真的追求。要想在繁杂的社会中不忘初心，只有家可以提供给你一个温暖的港湾。

绿色环保的肌理漆带来森林般的呼吸！麋鹿、野鹤、翠鸟与爱美的女神，在森林里舞蹈高唱，恍如空谷百合在雨后的彩虹下鲜嫩欲滴。迷失的小鹿带你寻找心中的乌托邦。

As the familiar saying goes, "East, west, home is best," which we quote from Frank Baum's The Wonderful Wizard of Oz. For the design of this project, the designer reveals his pursuit of warmth, texture and innocence. If you do not want to forget your original intention in the complex society, home is the only place that can give you a warm harbor.

The green and environmental paint brings some forest-like breath. The deer, wild crane, kingfisher and the beauty-loving goddess are dancing and singing in the forest, which seems like the valley lilies are blossoming under the rainbow after the rain. The missing deer would lead you to find the Utopia in your heart.

金地新外滩壹号样板间
Gemdale New Bund of No. 1, Show Flat

设计公司：杭州易和室内设计有限公司	Design Company: Hangzhou EHE Interior Design Co., Ltd.
设 计 师：李扬	Designer: Li Yang
软装设计公司：杭州极尚设计装饰工程有限公司	Soft Decoration Design Company: Hangzhou Jishang Design Decorative Engineering Co., Ltd.
软装设计师：祝竞如、江嘎、叶子丰	Soft Decoration Designers: Zhu Jingru, Jiang Ga, Ye Zifeng
项目地点：浙江宁波	Project Location: Ningbo of Zhejiang Province
项目面积：90 ㎡	Project Area: 90 m^2

这是一套温馨简洁的法式田园风格样板间，90㎡的面积经设计师精心设计后，整个空间充满明亮清新的气息。简约的法式风格中带着淡淡的田园气息，舍弃了奢华与铺张，角落里都是浓浓的小清新情调。线条流畅、雕花精致的家具、精心搭配的空间色彩、充满细节的摆设……浪漫小清新的风格中是无处不在的精致细节，让人沉浸在这如画的氛围中。

This is a warm and concise French pastoral style show flat. With the careful design of the designer, the 90m^2 area makes the whole space be filled with bright and fresh atmosphere. The concise French style carries some slight pastoral atmosphere, abandoning luxury and extravagance, while there are intensive fresh tones in the corners. Smooth lines, furniture of delicate carving, carefully collocated space colors and ornaments full of details… Inside the romantic and fresh style are delicate details everywhere, making people get immerged in the picturesque atmosphere.

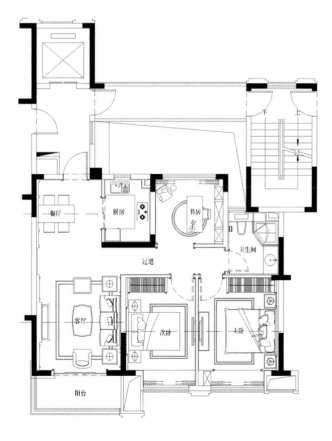

华林上景
Hualin Shangjing

设计公司：观云设计机构
设 计 师：林元君
项目地点：福建福州

Design Company: Guanyun Design
Designer: Lin Yuanjun
Project Location: Fuzhou of Fujian Province

怀旧与自然是贯穿整个设计的基调。清新脱俗的墙面颜色、自然朴实的木料及粗犷的手工陶砖为主要材质，为人们营造出一个慵懒舒适的室内空间。设计师在欧式乡村风格的基础上做减法处理，让整个空间显得年轻化，是适合广大年轻人的设计风格。

设计师将简洁实用的设计风格注入到本案的设计中来，并与现代的材质相结合，让古朴的自然气息穿透岁月，在我们的身边活色生香。

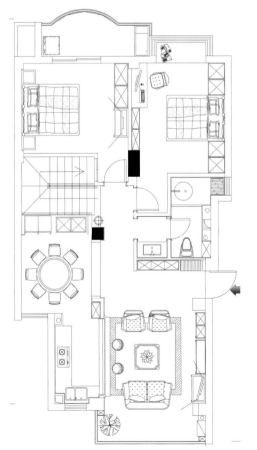
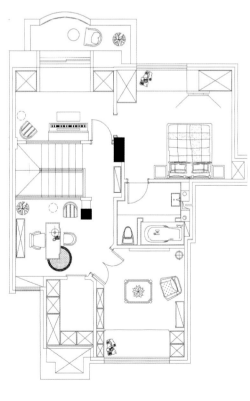

Nostalgia and nature are the keynotes running through the whole design. The major materials include refreshing wall, primitive wood materials and rough handmade earthenware bricks, producing a lazy and comfortable interior space for people. The designer applies subtraction treatment based on European pastoral style, making the whole space appear quite young, quite in accordance with the young people's requirements.

The designer instills concise and practical design style into the project design, which is combined with modern materials, making primitive natural atmosphere running through time, forming lively colors around us.

万科双月湾 TA2 户型
Vanke Lunas Del Mar, House Type TA2

摩登样板间 III 新田园

设计公司：深圳市昊泽空间设计有限公司
设 计 师：韩松
项目地点：广东惠州
项目面积：260 ㎡
主要材料：石材、瓷砖、木饰面、涂料

Design Company: Shenzhen Haoze Space Design Co., Ltd.
Designer: Han Song
Project Location: Huizhou of Guangdong Province
Project Area: 260 m²
Major Materials: Stone, Ceramic Tile, Wood Veneer, Coating

悠长假期

身未动，心已远。莫不如给自己来一次放逐，由身体到心灵，没有理由、没有借口、没有代价，放空身体里的贪嗔、邪恶、虚伪、聒噪，还原童真、质朴。空间之所触所及，没有技巧，不修边幅，只有勾起情绪的记忆就够了。

本案以最本色的面目还原出空间的质朴气质，它表达了主人不羁心灵下的质朴情怀。田园情调的设计风格与大自然亲密互动，落地窗、玻璃门外的空间延伸进来，将室内外连为一体。成就了居室主人返璞归真的理想生活情境。

MODERN SHOW FLAT III NEW PASTORAL 027

MODERN SHOW FLAT III NEW PASTORAL

A Long Vacation

Before you take off, your heart has already gone to a very far place. You'd better give yourself an exile, physically and psychologically, with no reason, no excuse, no cost emptying the anger, greed, viciousness, hypocrisy, and noisiness, and restoring innocence and plainness. For every detail of the space, there is no technique, no caring for the outlook, just being enough to arouse one's emotional memories.

This project uses the truest appearance to restore the space's primitive temperament, representing the property owner's plain emotions inside the uninhibited heart. The pastoral design style has intimate interactions with the grand nature, and the French window and space outside the glass door extend inside, connecting interior and exterior spaces as a whole, creating an ideal living situation for the property owner which can recover one's original simplicity.

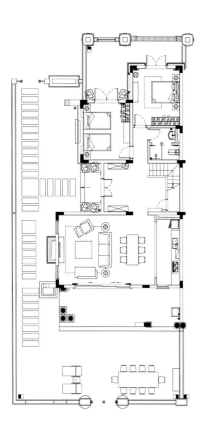
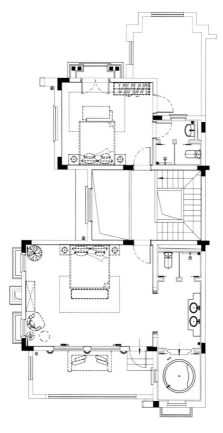

MODERN SHOW FLAT III NEW PASTORAL 031

万科双月湾 TB2 户型
Vanke Lunas Del Mar, House Type TB2

摩登样板间 III 新田园

设计公司：深圳市昊泽空间设计有限公司
设 计 师：韩松
项目地点：广东惠州
项目面积：300 ㎡
主要材料：石材、瓷砖、木饰面、涂料

Design Company: Shenzhen Haoze Space Design Co., Ltd.
Designer: Han Song
Project Location: Huizhou of Guangdong Province
Project Area: 300 m²
Major Materials: Stone, Ceramic Tile, Wood Veneer, Coating

本案强调空间的独立与联系，空间整体配色较为稳重、大气，注重营造唯美、典雅的气息，满足了业主对空间的要求，为业主打造出了与众不同的个性生活空间。设计师将新田园风格的精髓提炼出来，与硬装设计进行了完美的结合。无论是带给人轻松、自然感受的家具，还是精致的配饰，都被和谐地融合在一起，开放的设计理念更体现出业主独有的品味和对精致生活的追求。

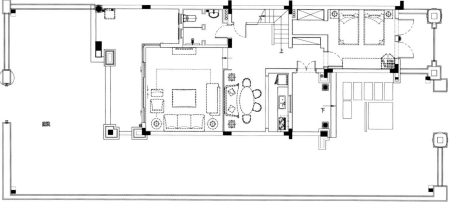

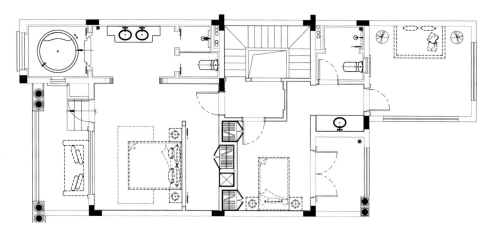

This project emphasizes the space's independence and connections. The space's whole color collocations are sedate and magnificent, focusing on the creation of aesthetic and elegant atmosphere, meeting with the property owner's requests for space, while producing a unique living space for him. The designer extracts the essence of new pastoral style, having perfect integration with hard decorative design. Either for furniture leaving people with relaxing and natural sensations, or delicate ornaments, all are combined harmoniously. The open design concept further represents the the property owner's peculiar tastes and pursuits of delicate life.

MODERN SHOW FLAT III NEW PASTORAL

摩登样板间 III
新田园

星雨华府
One Residence of Silver Rain Mansion

设计公司：北京元洲装饰南京分公司
设 计 师：陈琼
项目地点：江苏南京
项目面积：135 ㎡
主要材料：硅藻泥、瓷砖、壁纸、仿古瓷砖、实木柜体等
摄 影 师：金啸文

Design Company: Beijing Yuanzhou Decoration Nanjing Branch Company
Designer: Chenqiong
Project Location: Nanjing of Jiangsu Province
Main Materials: Diatom Mud, Ceramic Tile, Wallpaper, Archaized Tile, Solid Wood Cabinet
Photographer: Jin Xiaowen

本案开放式的空间结构、随处可见的花卉和饰品、雕刻精细的家具……所有的一切从整体上营造出一种法式地中海风格的浪漫气息，在任何一个角落，都能体会到主人悠然自得的生活和阳光般明媚的心情。

客厅：客厅墙面采用硅藻泥来涂饰，米色的墙体形成一种特殊的不规则肌理面。

餐厅：设计灵感来自南意大利和南法。南意大利向日葵的金黄和南法薰衣草的蓝紫相映，形成一种别有情调的色彩组合，具有自然田园般的美感。

This project has open style space structure, flowers and ornaments everywhere, carefully sculpted furniture... All these create in the whole some romantic atmosphere of French style Mediterranean style. No matter you are in which corner, you can always feel the leisurely life style and sunny moods of the property owner.

Living Room: The wall of the living room applies diatom ooze as the decoration and the beige wall forms some special irregular texture surface.

Dining Hall: The design inspirations arise from Italy and France. The gold yellow sunflowers from South Italy and royal purple of lavender from South France bring out the best in each other, producing some color combinations of peculiar emotions, with some natural pastoral aesthetic feels.

海口美林谷度假村
Haikou Meilin Valley Holiday Village

设计公司：深圳市鸿艺源建筑室内设计有限公司
设 计 师：徐静、郑鸿
项目地点：海南海口
项目面积：230 ㎡

Design Company: Shenzhen Hong Yi Yuan Architectural Interior Design Co., Ltd.
Designer: Xu Jing, Zheng Hong
Project Location: Haikou of Hainan Province
Project Area: 230 m²

主案设计师从度假的初衷出发，汲取东南亚风格的规整、朴实、精致，以法式田园风的自然、清新、浪漫为点缀，简单凝练的直线条铺垫出轻松愉悦的度假心情，原木家私与绿植掩映成趣，与自然共呼吸。因海岛湿热的气候，设计师特意在踢脚线处做了加高的防潮处理，而木纹砖拼接大理石的手法，更是完美地实现空间的划分。为了柔和硬装，设计师用异域味十足的软装进行二次美化，抽象粗犷的地毯挂画与清爽朴实的亚麻布艺，奏响了一支悠长而奇异的海岛舞曲。

进入空间，敦厚古典的泰式木雕为视觉重心，工笔雕刻的花叶，在两盏壁灯的映照下，枝舒叶卷，花开百态。

客厅中抽象纹理的地毯呼应着布艺沙发的淳朴与自然。水蓝色的藤蔓壁画将这一抹难以描状的色彩，蔓延到一边的垂帘，在八角明灯的浮光掠影里，咿咿呀呀地将一支渔舟小调唱出寻找未知的神奇力量。

圆镜勾勒出餐厅的温婉可爱,一字排开的吊灯是视线的焦点。敦厚温润的桌椅,默默不语却憨态可掬,仿若在听闻人们的趣闻笑语。

The leading designer starts from holiday and imports the orderly, austere and delicate qualities of South-East Asian styles, decorated with natural, fresh and romantic French style pastoral style and creating relaxing and pleasing holiday moods with simple and compact straight lines. The log wood furniture and green plants bring out the best in each other, echoing the natural environment. Due to the damp and hot climate of oceanic island, the designer specifically made some heightened dampproof treatment towards the skirting lines, while the approach of wood grain bricks matching marble perfectly achieves the space division. In order to soften the hard decorations, the designer made second-time beautification towards the soft decorations full of exotic feels. The abstract and robust carpet graphics and the primitive linen fabrics compose a lingering and peculiar oceanic island dance music.

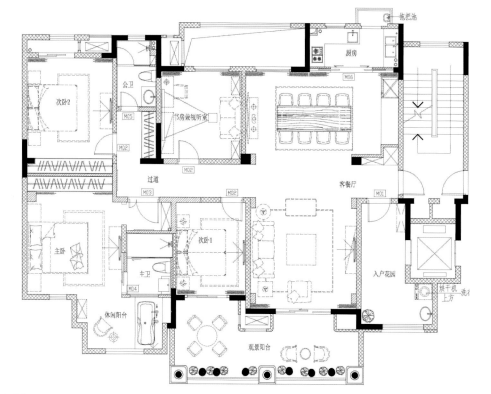

Upon entering the space, you can find the classical Thailand style wood carving as the visual focus. Setting off by two wall lamps, the floral leaves of fine brushwork carving display smooth leaves and branches, with blossoming and varying flowers.

For the living room, the carpet of abstract pattern echoes the primitive and natural cloth sofas. The water blue cirrus mural extends this inexplicable color to the curtain on one side. Among the light and shadow created by the octagonal light, the musical fisher-boat ditty presents some miraculous strength aiming for the unknown.

The round mirror makes the dining hall be mild and lovely, while the lined-up droplights become the visual focus. The smooth and mild tables and chairs are mute, yet charmingly naive, it is like that they are listening to people's interesting news and laughters.

让时间放慢脚步
Slowing down the Pace of Time

设计公司：福州大成室内设计有限公司
设 计 师：朱林海
项目面积：700 ㎡
主要材料：锈石荔枝面花岗石、日本伊奈千陶彩、仿古大理石、草编壁纸、透光板、抽丝仿古橡木、耐火砖、木纹灰仿古理石、铁艺、仿古手工砖等
摄 影 师：朱林海

Design Company: Fuzhou Dacheng Interior Design Co., Ltd.
Designer: Zhu Linhai
Project Area: 700 m²
Major Materials: Rust Stone Litchi Surface Granite, Japan INAX Colorful Earthenware, Archaized Marble, Straw Plaited Wallpaper, Board Pervious to Light, Wired Archaized Oak, Refractory Brick, Wood Grain Gray Archaized Marble, Artistic Iron, Archaized Hand-made Brick
Photographer: Zhu Linhai

此前已经给本案的业主设计了一套美式风格的别墅，拿到这个新案例的时候，我想我要带给他一种新的生活方式。考虑许久，想为他打造一个融合东方韵律与西方舒适性为一体的设计作品。在这里我通过大量的留白与粗犷的木材搭建一个主骨架，融合了东方的砖饰、竹子、自然面的石材与西方精致的铁艺、壁炉，再结合大量的自然光，打造了一个我心中的自由、轻松的空间氛围。

在这里随处是自然景观与室内情景的相互交错，仿佛在水中、在河边、在竹林中，模糊了室内与室外的界线，身处当中似时间放慢了脚步。

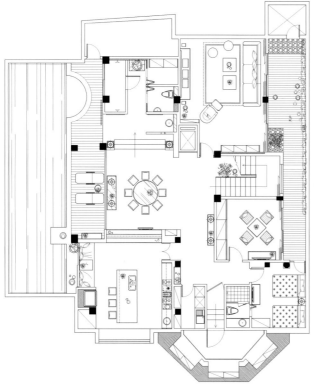

MODERN SHOW FLAT III NEW PASTORAL

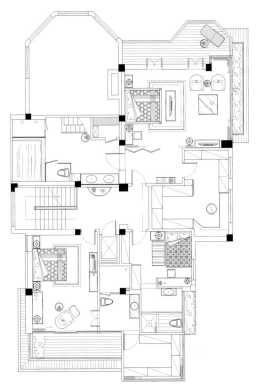
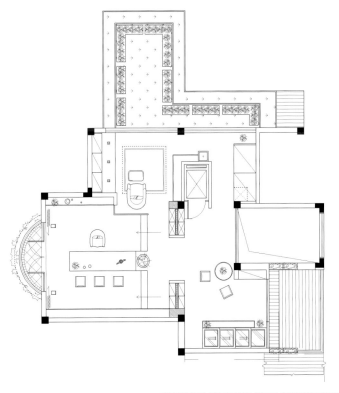

MODERN SHOW FLAT III NEW PASTORAL 055

Previously, I had designed an American style villa for the property owner. While I was commissioned with this new project, I intended to present to him a new life style. After much thoughts, I considered producing a design work for him which integrates oriental charms and western comfort. Here, I built a major skeleton through a large amount of blank spaces and wild materials, integrating oriental brick accessories, bamboos, stones of natural surface and delicate western iron art and furnaces, while combining a large amount of natural light and producing a free and relaxing space atmosphere.

Everywhere you can find the mutual intersection of natural landscape and interior sceneries, making you seem be at waterside, river bank or in the bamboo forest, which dims the boundaries between interior space and exterior space. And time here seems to slow its pace.

御湖半山
Royal Lake Mid-Levels

设计公司：方块空间	Design Company: Fang Kuai Space
设 计 师：蔡进盛	Designer: Cai Jinsheng
项目地点：江西南昌	Project Location: Nanchang of Jiangxi Province
摄 影 师：邓金泉	Photographer: Deng Jinquan

MODERN SHOW FLAT III NEW PASTORAL 061

本案为美式乡村风格，在设计上突出生活的舒适与自由，不论是感觉笨重的家具，还是带有岁月感的配饰，都在告诉人们这一点。布艺是乡村风格中非常重要的设计元素，本色的棉麻是主流，布艺的天然感与乡村风格能很好地协调起来。

美式乡村风格，它在古典中带有一点随意，摒弃了过多的繁琐与奢华，兼具古典主义的优美造型与新古典主义的功能配备，既简洁明快，又温暖舒适。美式家具有着简化的线条、粗犷的体积、自然的材质、较为含蓄保守的色彩及造型。但它以舒适为设计准则，每一件都透着阳光、青草、露珠的自然味道，仿佛随手拈来，毫不矫情。

This project has American countryside style, in design highlighting living comfort and freedom. Both the seemingly heavy furniture and the ornaments of time feel are telling people that. Fabrics is an important design element in countryside style, with natural color cotton and linen as the leading elements, while the natural feel of fabrics and countryside style can coordinate with each other greatly.

There is some casualty in classicism for American countryside style, abandoning too much complexity and luxury, while owning the elegant format of classicism and functional outfits of Neo-Classicism, which is concise and brisk, and warm and cozy. American style furniture have concise lines, rough scale, natural materials and comparatively restrained colors and formats. Yet it has comfort as the design guideline and each element is full of the natural tastes of sunshine, green grass and dewdrops, all appearing so natural, with no hypocritical feel at all.

MODERN SHOW FLAT III NEW PASTORAL 065

芬芳流淌 Flowing Fragrance

设计公司：福州大成室内设计有限公司
设 计 师：朱林海
主要材料：椿木、橡木、桦木、麻质地毯、仿古砖、英式壁纸

Design Company: Fuzhou Dacheng Interior Design Co., Ltd.
Designer: Zhu Linhai
Major Materials: Ailanthus Wood, Oak, Wood, Birch Wood, Fibre Carpet, Archaized Brick, English Wallpaper

本案寻找的感觉是时光沉淀的厚重感，大量搓色处理的饰面与古朴的色调，粗犷中有细腻感，很好地诠释了这个设计的主题。

客厅顶部的木质花饰有做旧的质感，仿佛是老房子遗留的装饰，精致的书架立在小巧别致的壁炉两侧，炉上有过圣诞节时还没有撤去的花环，上方墙面上有搜罗来的各种挂盘，呈现出主人热爱生活的情怀。

主卧粗犷的木饰背景墙，给人带来很强的安全感，通透的设计把卫生间、卧室与衣帽间贯穿起来，为生活带来了便利。

What this project searches for is the decorous feel accumulated by time. For the large amount of veneers through coating treatment and primitive color tones, there is the refined feel in the rough appearance, perfectly interpreting the theme of the design.

The wooden flower decoration on the ceiling of living room has antique feel texture, just like the decorations left from the old house. The delicate book shelves stand on both sides of the tiny and refined furnace, over which is a garland remained from last Christmas and on the wall above are the various kinds of collected hanging dishes, representing the owner's enthusiasm for life.

The rough wooden background wall of the master bedroom brings to people intensive safety feel, while the transparent design connects the wash room, master bedroom and cloakroom, creating some conveniences for life.

MODERN SHOW FLAT III NEW PASTORAL 071

摩登样板间 III 新田园

桂丹颐景园高层样板间
Guidan Yijingyuan High-Rise, Show Flat

设计公司：SDD 上达国际
　　　　　深圳市上达建筑装饰设计有限公司
设 计 师：刘海涛
项目地点：广东佛山
项目面积：90 ㎡

Design Company: SDD International
Shenzhen SDD Architectural Decoration and Design Co., Ltd.
Designer: Liu Haitao
Project Location: Foshan of Guangdong Province
Project Area: 90 m²

本案为英式田园风格，在整体色调上清新典雅。以象牙白或者奶白色为主题的天花板与家具显得既纯净又楚楚动人，再配上晶莹剔透的水晶吊灯，更使得房间从容而典雅。浅灰的地板，有一种古朴清新的感觉。室内陈设的家具或者饰品的色彩也是低调内敛的，没有喧宾夺主，但是又不可缺少。

象牙白的装饰，合着淡然的暖色灯光，看上去感觉很放松休闲。而无处不在的百合花，却也给厨房添加了几分生机。

This project has English pastoral style, with fresh and elegant whole color tone. The ceiling and furniture with ivory white and off-white color as the theme appear pure and attractive, accompanied with crystal droplights which make the space appear elegant and calm. The light gray floor board has some primitive and fresh sensations. The interior furniture and colors of ornaments are also low-profile and restrained, not stealing the show, while being a necessary part.

The ivory white furnishing decorations match the slight warm-color lights, making the whole space appear relaxing and leisurely. While the lily flowers everywhere add more vitalities to the kitchen.

摩登样板间 III
新田园

鸢尾花开
Blossoming Flower-de-luce

设计公司：花开设计工作室
设 计 师：林函丹
项目地点：福建福州
项目面积：118 ㎡

Design Company: Flower Blossom Design Studio
Designer: Lin Handan
Project Location: Fuzhou of Fujian Province
Project Area: 118 m²

本居室是业主准备退休后在此小憩、阅读、听音乐、与朋友聚会的场所。经过愉快的沟通，业主选中了美克美家的鸢尾乡旅系列的部分家具，设计师顺其自然地围绕这个系列家具的风格去展开空间的设计。

本案的最大特点就是搭配小众风格的家具必须由工匠现场手工打造，比如黑色亚光做旧的橱柜、主卧室中的黑色独立柜、米白色复古风格轻度做旧的木门、朴素的电视墙、高于常规高度并与家具木器漆颜色一致的踢脚线等，均由工匠现场制作。所以风格中少了工业化时代机械的味道，传递着充满休闲、精致、自然气息的小资情调。

This residence is for the property owner to take a rest, read some books, listen to the music, party with friends, etc., after he retires. Through happy negotiations, the designer selects some furniture of Markor Furnishing's Flower-deluce country trip series. The designer naturally centers on the style of this series of furniture to present the space design.

The biggest feature for the project is that furniture with minority style have to be made by craftsman on site, such as black matte surface ancient-style cabinets, black

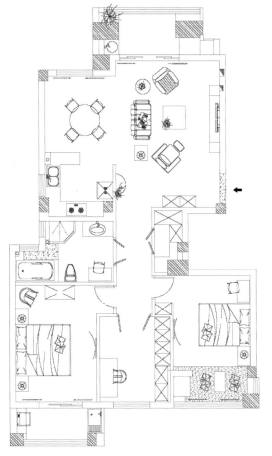

independent locker inside the master bedroom, beige white and slightly ancient wood door of vintage style, primitive TV background wall, skirting lines higher than normal height yet with consistent color with the furniture wood lacquer. Thus inside the style, there is not that much taste of machines of industrial age, sending out some bourgeoisie tones of leisurely, delicate and natural atmosphere.

城市花园 City Garden

设计公司：张之鸿空间设计工作室
设 计 师：张之鸿、孙滔
施工单位：常熟市创亿装饰工程有限公司
项目地点：江苏常熟
项目面积：145 ㎡
主要材料：珠光涂料、木地板、大理石砖等

Design Company: Zhang Zhihong Space Design Studio
Designers: Zhang Zhihong, Sun Tao
Construction Company: Changshu Chuangyi Decoration Engineering Co., Ltd.
Project Location: Changshu of Jiangsu Province
Project Area: 145 m²
Major Materials: Pearl Luster Coating, Wood Floor, Marble Bricks

法式乡村风格将以人为本、尊重自然的传统思想作为设计理念，使用令人备感亲切的设计元素，创造出如沐春风般的感官效果，属于自然风格系列。

本案的法式田园之家少了一点美式田园的粗犷与英式田园的厚重、浓烈，多了一点大自然的清新与浪漫……

根据业主的生活习惯，在145 ㎡的空间里做成了中西双厨，西厨和餐厅、客厅没有明显的分隔，使空间看起来更大，每个卧室都设置了单独的衣帽间。

卧室设计重视氛围的营造,纯正法式风格中繁复的雕花设计不适合现代人的审美观,因此,设计师去掉了法式风格中的这些复杂元素,保留了一些清新的设计元素。比如彩色的墙面、花朵、布艺和装饰画的大量运用等,对卧室氛围的营造具有很好的烘托作用。

French pastoral style has as design concept human-oriented traditional thinking with respect for nature, makes use of amiable design elements and creates sensory effects like spring breeze, which belongs to be natural style.

The French pastoral style home of this project lacks a bit of ruggedness of American style countryside and the heaviness and intensity of British countryside, but with the freshness and romantic feel of grand nature...

According to the life habits of the property

owner, the designer creates China and western style kitchens inside this 145m² space. There is no distinct separation between western kitchen, dining hall and the living room, which makes the space appear larger, and each bedroom is set with an independent cloakroom.

The bedroom design focuses on the creation of atmosphere. And the complicated carving design of pure French style is not fit for modern people's aesthetic standards. Thus the designer removes these complex elements in the French style and retains some fresh design elements. The many applications of colorful wall, flowers, fabrics and decorative paintings play an important role in the creation of bedroom atmosphere.

摩登样板间 III
新田园

游于艺的 Feeling
Feelings Swimming in Art

设计公司：老鬼设计事务所
设 计 师：杨旭光
项目地点：广东深圳
项目面积：533 ㎡

Design Company: Laogui Design Firm
Designer: Yang Xuguang
Project Location: Shenzhen of Guangdong Province
Project Area: 533 m²

本案以最朴实的手法，最真实的心情抒发自己的情感，告知我们"游于艺"的最高境界。设计师主张以生活情趣来点缀空间的每一个细节，无形中透露出主人对空间的主人翁的意识。客厅铺贴的色调保持与家具的色系一致，采用灵活铺贴的手法，使整个空间不过于单一。在软装饰的选择上，保持与整个空间的风格相协调，自然而不拘谨。

在楼梯的处理手法中，用先"破"后立的手法，用具有民族风格图案的瓷砖来点缀每一个踏步，侧面的灯光的衬托，以及楼梯扶手的"破"，使楼梯的设计成为空间的亮点，完全符合"游于艺的 Feeling"的主题设计。

This project displays the emotions with most primitive approaches and most sincere moods, telling us the highest state of "Feelings Swimming in Art." The designer advocates decorating each detail of the space with life interests, invisibly revealing the property owner's sense of ownership towards the space. The color tone of living room maintains similarity with the color system of furniture, with dynamic pavement approaches, which makes the whole space do not appear monotonous. The selection of soft decorations maintains harmony

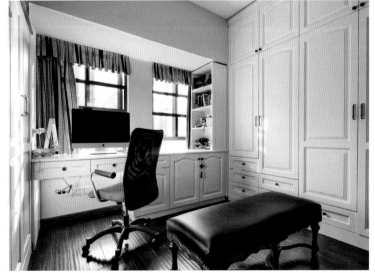

with the style of the whole space, being natural and dynamic.

Af for the treatment of staircase, the designer makes use of the approaches of making the staircase become the highlight of the space, through decorating each step with ceramic tiles of national style graphics, setting off by siding lightings and the handrails of staircases, totally in accordance with the thematic design of "Feelings Swimming in Art."

MODERN SHOW FLAT III NEW PASTORAL

半岛泉水欣座3号

Peninsula·Quanshui Xinzuo, Building No. 3

设计公司：大连非常饰界设计装饰工程有限公司
设计师：马燕艳
项目面积：90 ㎡

Design Company: Dalian Feichang Shijie Design Decorative Engineering Co., Ltd.
Designer: Ma Yanyan
Project Area: 90 m²

本样板间的设计倡导"回归自然"，美学上推崇"自然美"，认为只有崇尚自然、结合自然，才能在当今快节奏的社会生活中获取生理和心理的平衡。家具多选用奶白、象牙白等色调，细致的线条和高档油漆的处理，使得每一件产品就像优雅成熟的女子含蓄温婉、内敛而不张扬，散发着从容淡雅的生活气息。

客厅搭配了两个造型不同的休闲椅和小圆几，闲时品一杯香浓的咖啡，想必是一个不错的选择。造型优雅的田园台灯是必不可少的配角。布艺的选择上多以田园碎花的图案为主，可以更好地营造出田园的气息。

The design of the show flat advocates "back to nature," and "natural beauty" aesthetically. The designer believes that only respect for nature and combination of nature can attain physical and psychological balance in modern social life of fast pace. The furniture mainly selects color tones of ivory white and milky white, with treatment of delicate lines and high-end paint, which make each product appear like an elegant and

mature woman with restrained, low-profile and gentle characteristics, sending off mild and graceful life atmosphere.

The living room is collocated with a leisure chair and a little round table of different formats, while you are free, you can enjoy a cup of strong coffee here. What a good choice! Pastoral table lamp of graceful format is a necessary supporting role here. The fabrics mainly focus on graphics of pastoral floral pattern, which could well create the pastoral atmosphere.

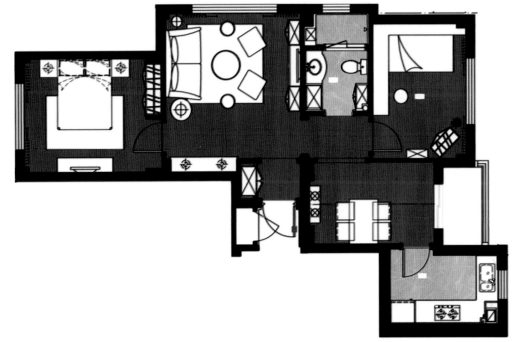

MODERN SHOW FLAT III NEW PASTORAL

中冶尚园 37 幢
MCC Shangyuan, Building No. 37

设计公司：大墅尚品·由伟壮设计	Design Company: Dashu Shangpin•Zhuang Design
设 计 师：由伟壮	Designer: You Weizhuang
软装设计：翁布里亚专业软装机构	Soft Decoration Designer: Umbria Professional Soft Decoration
项目地点：江苏常熟	Project Location: Changshu of Jiangsu Province
项目面积：145 ㎡	Project Area: 145 m^2
主要材料：仿古砖、橡木地板、壁纸、护墙板	Major Materials: Archaized Brick, Oak Wood Floor, Wallpaper, Wainscot Board

本案定位为美式乡村风格设计,摒弃了繁琐与奢华,以舒适机能为导向,强调"回归自然",使家居空间变得更加轻松、舒适。在结构造型上相对简单,色彩朴素,突出了生活的舒适与自由,不论是感觉笨重的家具,还是带有岁月沧桑的配饰,都散发出休闲、怀旧、自然的气息。

生活是人生的片段,居所是都市的一隅。出则自然,入则繁华,居所设计所营造出的文化感、贵气感、沉静感,是居住于都市一隅的每一个人都心生向往的。

This project is oriented to be American pastoral style, abandoning complexity and luxury, guided by comfort functions, emphasizing "returning to nature," and making furnishing space become more relaxing and comfortable. The structural format is comparatively simple, with plain colors, highlighting the comfort and freedom of life. Both the seemingly heavy furniture and the ornaments of time traits sent out some leisurely, nostalgic and natural atmosphere.

Daily life is a part of whole life, and residence is a corner of city. Stepping outside, there is the nature, moving inside, there is the prosperity. The cultural feel, noble fell and tranquil feel created by the design are what every people living in a corner of city longs for sincerely.

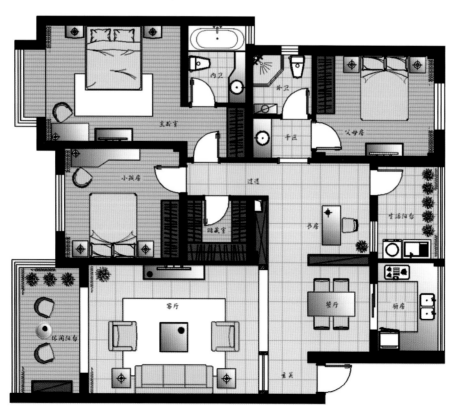

东湖京华京玉苑
East Lake Jinghua Jingyu Garden

设计公司：大墅尚品·由伟壮设计
设 计 师：由伟壮
项目地点：江苏常熟
主要材料：仿古砖、地板、饰面板、壁纸、石膏线条

Design Company: Dashu Shangpin·Zhuang Design
Designer: You Weizhuang
Project Location: Changshu of Jiangsu Province
Major Materials: Archaized Brick, Floor Board, Veneer Board, Wallpaper, Plaster Lining

客厅、餐厅： 客厅与餐厅的设计使用美式风格中常见的勾缝处理方法，线条圆润优美、结构简洁大方。空间墙面以浅黄色墙为主，交错使用油画挂件等进行搭配，空间色彩丰富、氛围温馨。棉麻质地的沙发，宽松柔软，十分舒适。古朴质感的茶几与桌椅等，展现出美式风格的自然气息。

厨房： 厨房顶面采用铝扣板集成吊顶，橱柜选用大理石台面的整体橱柜，门板柜体采用美国红橡材质。灶台上方做了一排吊柜，增加了储物功能。

卧室： 主卧墙面采用浅绿色壁纸为主，呈现出浪漫、沉静的生活氛围。浅色的天花与墙面将淡绿的床头背景包围起来。实木制作的床具、柜体等完美地呈现出美式家具的稳重与精美。

Living Room and Dining Hall: The design of living room and dining hall applies the jointing treatment approach quite common in American style, with smooth and graceful lines, and concise and grand structure. The walls focus on light light yellow color, collocated with staggered oil paintings and pendants. And the space has abundant colors and warm atmosphere. The sofa of linen and cotton texture is quite soft and comfortable. The tea table and chairs of primitive texture display natural atmosphere of American style.

Kitchen: The kitchen ceiling applies pinch plate accumulated ceiling, the cabinets select integrated cupboard of marble surface and the door plank has American red oak materials. There are a set of wall cupboards over the hearth, adding to the storage functions.

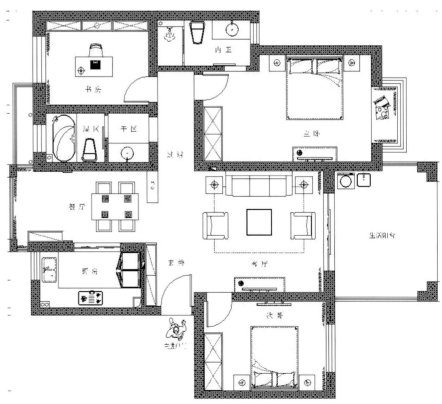

Bedroom: The wall of master bedroom applies light green color wallpaper, displaying some romantic and tranquil life atmosphere. The light color ceiling and wall enclose the light green bedside background. The solid wood bedding accessories and cabinets perfectly display the sedateness and delicacy of American style furniture.

雅居乐·情蓄向日葵
Agile Properties·Love of Sunflowers

设计公司：昶卓设计·黄莉工作室
软装设计：昶卓软饰
施工单位：怡明施工
项目面积：87 m²

Design Company: Changzhuo Design·Huang Li's Studio
Soft Decoration Designer: Changzhuo Soft Decoration
Construction Company: Yiming Construction
Project Area: 87 m²

本案的设计是业主对于自己想要的"家"的完整诠释，这里有他们前往希腊看到的阳光、大海，这里有他们性格里向日葵温暖的明黄。

门厅处，设计师将原入户花园进行了改造，做了一个储物间，同时设计了一个卡座，方便人们进门放包或者是在上面坐着换鞋。

客厅的窗帘选择了竖向的条纹，延展了空间的高度，同时也与沙发的竖条纹相呼应。电视背景墙的尺寸原本并不充裕，刻意将过道的梁做了弧形的收口，将视线延展了不少。墙面明亮的黄色与海洋的蓝色系沙发搭配在一起，似乎让人能感受到阳光、沙滩的味道。

餐厅中，原来入户花园的墙拆除后，并不是直接砌了墙，而是将之利用起来做了一面酒柜，可以摆放喜欢的饰品，也可以放上家人的照片。厨房的门不是传统的移门，而是做了一个隔断，使两个空间的互动性增加了不少。

The project design is the perfect interpretation for the "home" that the property owners want. There are the sunshine, the ocean that they say in Greece. And there are the warm bright yellow colors of sunflowers in their characteristics.

In the hallway, the designer made some changes towards the indoor garden and set a storage space here, with a seat which can hold people's bags or for people to sit while changing shoes.

The curtain of living room selects vertical stripes, extending the height of the space, while echoing the vertical stripes of the sofa. The scale of TV background wall was not quite sufficient originally, the designer deliberately made an arch closing towards the beam of corridor, greatly extending the lines of sight. The bright yellow color of the wall is collocated with the oceanic blue color sofa, making people seem to feel the taste of sunshine and beach.

For the dining hall, after the indoor garden's wall was removed, the designer did not build a wall directly, but made use of the space to make a wine cabinet to put the ornaments that the property owner likes, or the family photos. The kitchen door is not the traditional sliding door, but a partition, adding to the interaction between the spaces.

书香绿苑·钟鼎山庄

Scholarly Green Garden · Zhongding Mountain Villa

设计公司：昶卓设计·黄莉工作室
软饰设计：昶卓软饰
项目面积：129 ㎡

Design Company: Changzhuo Design•Huang Li's Studio
Soft Decoration Design: Changzhuo Soft Decorations
Project Area: 129 m²

历代的贵族雅士，总是会给自己的住宅起个名字，贵族附庸风雅，士人用以励志，他们多半会客在客厅，会友在书房。现代的装修，业主还能有此雅兴的并不多。女主人是个集优雅、博学于一身的知性女子，为这套住宅命名为"书香绿苑"是再合适不过的了。

客厅中满墙的书成了这套房子最大的亮点，电视墙这边就没有必要做任何刻意的装饰。从餐厅看向客厅，你会发现顶上的壁纸与酒柜上的壁纸遥相呼应。餐厅的壁纸从顶面延伸到墙面，拉伸了空间，吊扇灯既美观又实用。

因为业主的工作需要，希望有一处书房，因此我们将阳台做成了书房，书柜与客厅沙发后的书柜形式统一起来。

主卧室的床头背景墙上是一幅专门定制的淡绿色的花鸟图，窗帘也用同样的淡绿色，清雅的感觉扑面而来。

MODERN SHOW FLAT III NEW PASTORAL

The nobilities and refined scholars of the past dynasties would like to give their residences a name. The nobilities mingled with men of letters and posed as a lover of culture. The scholars applied the name as an encouragement. They usually received guests in the living room, and met friends in the study. With modern decorations, there are few property owners inheriting that aesthetic mood. The hostess is a refined woman with abundant elegance and learnings. It is quite appropriate to name this residence "Scholarly Green Garden."

The whole wall books of the living room is the biggest highlight of this house. It would be unnecessary to make any deliberate decorations on the TV background wall. Looking to the living room from the dining hall, you would find that the ceiling wallpaper and the wallpaper of the wine cabinet are echoing each other. The wallpaper of dining hall extends from ceiling to the wall, extending the space. The ceiling fan light is nice-looking and practical.

Due to work requirements, the property owner needs a study. Thus we make the balcony into a study. The book cabinet and the one behind the sofa in the living room is coordinated with each other in format.

The bedside background wall of the master bedroom is a custom-made light green painting of flowers and birds in traditional Chinese style. The curtain also applies the same light green color, producing fresh and elegant atmosphere.

MODERN SHOW FLAT III NEW PASTORAL

日湖花园某住宅
One Residence of Rihu Garden

设计公司：任朝峰设计事务所
设 计 师：任朝峰
参与设计：刘鹏、施海杰、蔡其辉
项目面积：145 ㎡

Design Company: Ren Zhaofeng Design Studio
Designer: Ren Zhaofeng
Associate Designers: Liu Peng, Shi Haijie, Cai Qihui
Project Area: 145 m²

在这个不大的空间里，有太多的内容需要添加进去，进去的时候会有冲突，会有不和谐的因素。设计师更多的工作就是把这些不和谐的、冲突的因素使之协调起来。一个家庭不光需要血缘的关系，还需要爱、需要关怀、包容。设计师把每一个空间元素想象，当然这些设计元素都有共同的文化背景，这样能够很好地协调与融合这些设计元素，能够把冲突的、不和谐的因素规避掉。

对于住宅设计来说，最难的就是完成度的问题。未完成有时候不一定是一件不好的事情，因为它应该是一个可以持续设计的过程。

Inside this space not too big, there are too many contents that need to be added inside. In the beginning, there might be conflicts and discordant elements. The work of the designer is to coordinate these discordant and conflicting elements. A family not only needs blood relations, but also needs love, cares and tolerance. The designer sincerely thinks about each space element, and of course these design elements have the common cultural background, which can finely coordinate and integrate these design elements, while avoiding conflicting and disharmonious elements.

As for residential design, the hardest part is the issue of completion. Sometimes, incompletion is not a bad thing, as it should be a process of sustainable design.

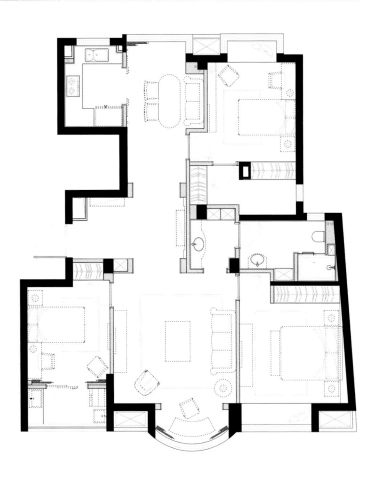

半岛时光·城市花园 22 幢
Peninsula Time·City Garden, Building No. 22

设计公司：大墅尚品·由伟壮设计	Design Company: Dashu Shangpin·Zhuang Design
设 计 师：由伟壮、张喜波	Designers: You Weizhuang, Zhang Xibo
施工单位：大墅施工	Construction Company: Dashu Construction
软装设计：翁布里亚专业软装机构	Soft Decoration Designer: Umbria Professional Soft Decoration Institution
项目地点：江苏常熟	Project Location: Changshu of Jiangsu Province
项目面积：130 ㎡	Project Area: 130 m²
主要材料：饰面板、硅藻泥、马赛克、墙地砖、地板、大理石	Major Materials: Veneer, Diatom Ooze, Mosaic Tile, Wall Floor Tile, Floorboard, Marble

柔美雅致的线条、遒劲有力的节奏感与有条不紊的曲线融为一体。米黄色的基调混搭白色和蓝色的家具，素色简约的条形沙发被花色缤纷的抱枕所点缀，铁艺的田园灯又成为这个暖暖的、情意浓浓的空间中的点睛之笔。

本案力求表现舒畅、自然的田园生活情趣，营造一个幽静休闲、轻松舒适、健康环保的家。即使足不出户也能时刻感受大自然的清新与美好。打开门，心情就如春天般灿烂……

Soft and elegant lines, strong rhythmic feel and orderly curves integrate as a whole. Beige yellow tone color is mingled with white and blue furniture, plain color bar-stripe sofa of concise style is ornamented with colorful bolster and iron pastoral lamps make the finishing point for this warm space of profound tender feelings.

This project tries the best to produce relaxing and natural pastoral life interests, thus creating a home which is tranquil, leisurely, comfortable, healthy and environmental. Even if you do not step outside, you can feel the freshness and beauty of nature all the time. Opening the door, the feelings are just so great like being in spring...

摩登样板间 III
新田园

世茂外滩新城
Shimao the Bund New City

设计公司：北岩设计
设 计 师：于园
项目面积：120 ㎡
主要材料：硅藻泥、仿古砖、木饰面等

本案为美式风格的田园设计，主要的居住对象是业主的父母。业主有时间回来陪父母居住，因此，设计师只保留两个卧室，尽量扩大空间的视觉感。拆除原厨房，将小阳台收纳进来，再一分为二，合并成一个中厨、一个西厨。餐厅的视线开阔了，厨房也被放大。所有的材质也尽量简化，让空间呈现最强的舒适性。

Design Company: Beiyan Design
Designer: Yu Yuan
Project Area: 120 m²
Major Materials: Diatom Ooze, Archaized Brick, Wood Veneer

This project has American style pastoral design, with the property owner's parents as the primary inhabitants. Sometimes, the property owner would come to live with his parents. Thus, the designer only reserves two bedrooms, maximizing the visual feel of the space. Removing the original kitchen, the designer incorporated the little balcony inside, splitting it into two parts and combining into a Chinese kitchen and a wester kitchen. The sightlines of the living room are broadened, and the kitchen is also expanded. All the materials are simplified to the greatest extent, making the space present the most powerful comfort feel.

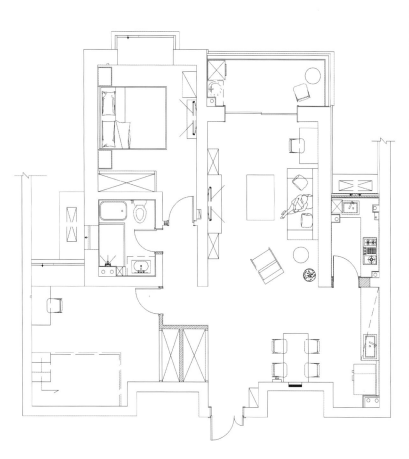

金茂四季花园 12 幢

Jinmao Four Season Garden Building 12

设计公司：大墅尚品·由伟壮设计	Design Company: Dashu Shangpin•Zhuang Design
设 计 师：由伟壮、李健	Designers: You Weizhuang, Li Jian
施工单位：大墅施工	Construction Company: Dashu Construction
软装设计：翁布里亚专业软装机构	Soft Decoration Designer: Umbria Professional Soft Decoration Institution
项目地点：江苏常熟	Project Location: Changshu of Jiangsu Province
项目面积：160 ㎡	Project Area: 160 m²
主要材料：实木复合地板、有色涂料、壁纸、仿古砖、铁艺等	Major Materials: Parquet, Colored Coating, Wallpaper, Archaized Brick, Iron Art

设计风格的创新点： 家不再是由苍白冰冷的墙面和死气沉沉的地面组成的，给家一个色调，给家一种精彩。在复古风中寻找那些遗失的美好，发掘出别样的精彩是本案设计的宗旨。

空间布局设计上的创新点： 空间本身较为方正，厨卫空间略显凌乱，设计师重新改造了卫生间中马桶旁边的墙体，将干区外置，增加了马桶旁边空间的面积。厨房面积比较小，外面靠墙部位增加了一排橱柜，既美观又实用。

设计选材上的创新点： 以暖色调为主，复古图案的米色壁纸、棕木饰面隔板，与复古浮雕面板装点的电视墙相对应。餐厅中古铜色的美式造型吊灯下，同色调的壁纸和木饰面相互映衬。客厅电视柜仿古花架上插着紫色薰衣草的花束，旁边放置着美式风格书架，装饰台灯与小鸟的饰品相结合，这样的搭配惬意而美好，凸显出业主热爱生活、富有生活情调的心境。

Innovation Points of Design Style: Home is no longer composed of pale cold wall and inanimate ground. It is the design tenet for this project to produce a tone and a spirit for home, and search for the lost goodliness in the antique style.

Innovation Points of Space Layout Design: The space is kind of square, with a bit messy kitchen and bathroom space, thus the designer regenerates the wall along the closestool within the bathroom, puts the dry area outside and adds the area of space along the closestool. The area of kitchen is kind of little, with a row of cabinets along the wall outside, being nice and practical at the same time.

Innovation Points of Materials Selection: The space focuses on warm color tone, with beige wallpaper of antique graphics and brown wood veneer clapboard echoing wallpaper and wood veneer of same color tone. There is a bouquet of purple lavender on the archaized pergola of the living room's TV cabinet, with an American style book shelf in the near location. The decorative lamps are combined with the little bird decorative objects, being pleasing and nice, and highlighting the property owner's mental states with passion for life and full of life emotional appeals.

MODERN SHOW FLAT III **NEW PASTORAL**

加州阳光
Sunshine of California

设计公司：大品装饰 | Dolong 设计
设 计 师：颜旭
项目地点：江苏南京
项目面积：130 ㎡
主要材料：仿古艺术砖、大理石、进口壁纸、进口地板等

Design Company: Dapin Decoration | Dolong Design
Designer: Yan Xu
Project Location: Nanjing of Jiangsu Province
Project Area: 130 m²
Major Materials: Archaized Artistic Brick, Marble, Imported Wallpaper, Imported Floorboard

本案以典雅的欧式风格为基调，素雅的色彩搭配深色木料，沉稳内敛之中流露出些许柔情，仿佛经历过时光的打磨和岁月的沉淀之后呈现在人们面前，这是一种与世无争的静谧心态。

推开门，可以感觉到这个空间的优雅与热情。房子的颜色搭配得很清新：浅色的美克美家家具、深色的原木地板、淡蓝色的墙面，散发出淡淡的典雅气息。客厅里简洁雅致的柱体将墙面划分为几段，黑、白、灰马赛克装点的地台成就了一个陈列台，既可以摆放小型绿色盆栽，也可以存放音响设备，仿古艺术砖贴面的电视墙在两个柱体之间形成独特的风景，大气、明朗。布艺沙发在色彩上与木地板、电视背景墙协调一致，整个空间不显陈旧，只留有岁月沉淀的韵味，历久弥香。

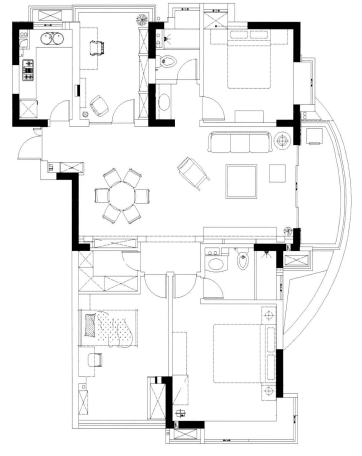

MODERN SHOW FLAT III NEW PASTORAL

This project has elegant Europeans style as the tone, with plain colors collocated with dark wood materials, displaying some gentle feelings in the sedate and profound atmosphere, which is presented in front of people like going through honing and passing of time. That is some tranquil state of mind standing aloof from worldly success.

Upon opening the door, you can find the elegance and ardor of the space. The collocation of the house is quite refreshing, light color Markor Furnishings, dark color log floor board, and light blue wall sending out some slight elegant atmosphere. Inside the living room, the concise and delicate pillars divide the wall into several sections. The platform decorated with black, white and gray mosaic tiles produces a presentation platform, which can hold little green potted plants and stereo equipments. The TV wall of archaized

artistic brick veneer produces some peculiar scenes between the pillars, being bright and magnificent. The fabric sofas coordinate with the wood floor board and TV wall in colors. The whole space does not appear antique at all, only with the lingering charms of the passed time.

摩登样板间 III 新田园

金海湾某宅
One Residence of Golden Bay

设计公司：深圳市伊派室内设计有限公司
设 计 师：段文娟
项目地点：广东深圳

Design Company: Shenzhen Yipai Decoration Co.
Designer: Duan Wenjuan
Project Location: Shenzhen of Guangdong Province

家是日积月累的印记,流溢满屋的温馨。本案业主喜欢自然、舒适的设计风格,空间不需要过于华丽及非常刻意的装饰。喜欢有细节的装点,希望家里能洋溢着春天的气息。所以,舒适自然的田园风格最能表达业主的喜好。

客厅中,质朴的格子沙发显得低调、素雅,电视背景墙上太阳花形的装饰物为室内增添了些许调皮的氛围,随处装点的小饰物都是那么可爱有趣。整个室内空间都用木地板来铺设,木色调渲染的氛围与自然田园风格的室内氛围相协调。

Home are the traits accumulated over a long period and warmth all over the house. The property owner of this project likes natural and cozy design style and the space does not have much too magnificent or quite deliberate decorations. The owner likes ornaments with details and hopes that home is filled with spring atmosphere. Thus, cozy and natural pastoral can best express the property owner's likes.

For the living room, the plain lattice sofa appears low-profile and elegant and the decorative objects on the TV background wall of sunflower format produce some mischievous atmosphere for the interior space, while the little ornaments everywhere all appear so lovely and interesting. The interior space applies wood floorboard as the pavement, and the wood color atmosphere and natural pastoral style interior atmosphere coordinate with each other.

金陵尚府
Jinling Capital Metropolis

设计公司：东易日盛南京分公司	Design Company: Dong Yi Ri Sheng's Nanjing Branch
设 计 师：陈熠	Designer: Chen Yi
项目地点：江苏南京	Project Location: Nanjing of Jiangsu Province
项目面积：135 ㎡	Project Area: 135 m²
摄 影 师：金啸文	Photographer: Jin Xiaowen

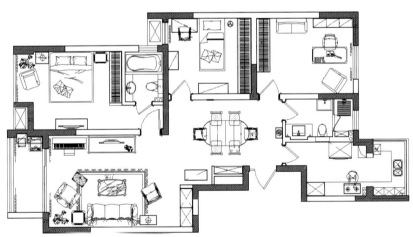

生活若是有味道，那该是一种馨香；生活若是有姿态，那该是一种优雅；生活若是有风景，那该是一种田园。本案的美式乡村田园风格正迎合了这一理念，古典中透露出恬淡、浪漫和自然的生活气息。

设计师在天花造型和家具的选择上以美式风格独有的花纹和雕饰来凸显典雅淳朴的气质。另外，材质的选用和色彩搭配上的创新也很好地体现出空间的文化底蕴，加上布艺与壁纸的完美结合，更体现出设计师对色彩搭配的娴熟把控。

If life has a smell, it shall be fragrance. If life has a posture, it shall be elegant. If life has a scene, it shall be the pastorale. For this project, the American countryside pastoral style meets with this concept, displaying some tranquil, romantic and natural life atmosphere in the classical atmosphere.

As for the selection of ceiling format and furniture, the designer makes use of patterns and carvings exclusive to American style to highlight the elegant and primitive temperament. Other than that, the materials selection and color collocations innovation finely represent the space's cultural deposits. Together with the perfect integration with fabrics and wallpaper, the project further displays the designer's mature mastering of color collocations.

卷珠帘 Bead Curtains

设计公司：一空设计事务所
设 计 师：沈一
项目面积：280 ㎡
主要材料：有色乳胶漆、拼花地板、孔雀仿古砖、进口壁纸

Design Company: YKON Design Studio
Designer: Shen Yi
Project Area: 280 m^2
Major Materials: Colored Emulsion Paint, Block Floor, Peacock Archaized Brick, Imported Wallpaper

美式的空间融入中式元素，但是却点到为止。用这种混搭的手法来塑造出具有中式文化底蕴的田园生活空间是本案的特点。黄绿的色调为整体空间营造出明快、活跃，犹如春天般的气息。

不同形态的仿古地砖划分出了功能空间的界限；文化砖打造的电视背景墙显得粗犷而又充满美感；楼梯墙上错落有致的生活照是一道别致的风景，有一种独属于家的味道。

The American style space is integrated with Chinese elements, yet with proper amount. This mix and match approach produces the pastoral life space of Chinese cultural connotations, which is the feature of the project. Yellow and green color tone makes the whole space appear brisk and dynamic, just like spring.

The archaized bricks of different formats produce boundaries of functional spaces. The TV background wall of cultural bricks appear grand and full of aesthetic feel. The well-proportioned life photos on the staircase wall is a peculiar scene, with a taste exclusive to home.

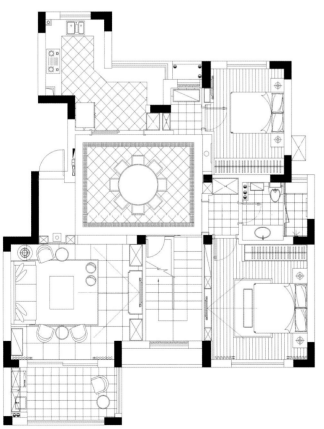
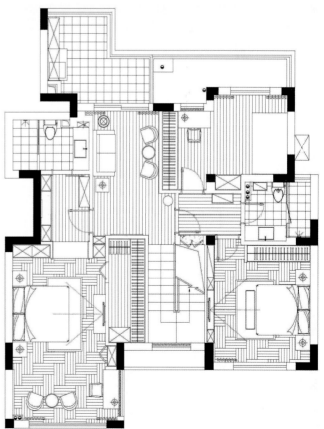

水岸风情
Waterfront Charms

设计公司：花开设计工作室
设 计 师：林函丹
项目地点：福建福州
项目面积：110 ㎡

Design Company: Flower Blossom Design Studio
Designer: Lin Handan
Project Location: Fuzhou of Fujian Province
Project Area: 110 m²

本案的整体风格是一种简约与田园的碰撞，与车水马龙的都市相比这里不够奢华，却体现出休闲放松的气息。展现了居室主人人淡如菊、平和淡定的心态。

整体空间浓淡相宜的木色调呈现出一种质朴的空间情感，偶然间点缀其间的紫罗兰色、青苹果色、橘色又为这份淡定的环境渲染出一丝朝气，一份温馨，冷暖色调的配合搭配出充满诗意的家居生活场景。

The whole style of this project is the clashing of conciseness and pastoral style. Compared with the city full of hustle and bustle, it is not quite luxurious here, yet with some leisurely and relaxing atmosphere, presenting the tranquil and peaceful state of mind of the property owner.

The wood color tone of appropriate shades for the whole space displays some primitive space emotions, decorated with scattered violet, green apple and orange colors, which create some optimistic and warm feel for the peaceful environment. The collocations of warm and cold color tones produce the poetic home furnishing scenes.

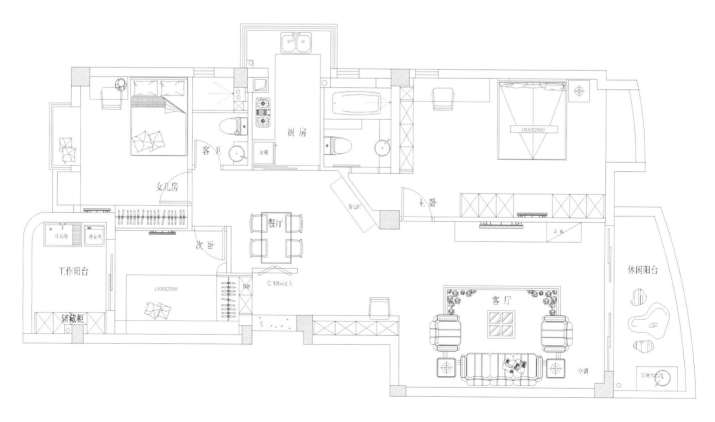

摩登样板间 III
新田园

细软时光
Refined and Mild Time

设计公司：花开设计工作室
设 计 师：林函丹
项目地点：福建福州
项目面积：110 ㎡
主要材料：瓷砖、金刚板、芬琳漆

Design Company: Flower Blossom Design Studio
Designer: Lin Handan
Project Location: Fuzhou of Fujian Province
Project Area: 110 m²
Major Materials: Ceramic Tile, Laminate Flooring, Feelings Paint

屋子里有一种气味，
像花园。
我很喜欢，
在那里待一下，
阳光从另一个房间进来，
说远不远的他方，
热与爱穿过窗缝。
收拢了流离的爱，
失所的梦。
阳台上的绿芽调皮地画个点，
面向天空，
弯腰、扭身、伸展枝叶，
在这个世界有氧运动着，
邀约你，来！
以梦养颜。
即便雨声也总文文静静，
答答滴滴。

There is some smell inside the house,
Which smells like garden.
I like that very much,
And stay there for a moment.

Sunshine gets inside from another room,
Which is not that far away.
Heat and love go through the windows,
Collecting the lost love and lost dreams.

The green buds on the balcony draw a bud mischievously,
Facing the sky,
Stooping, twisting, and spreading the leaves and branches,
Doing some aerobic exercises,

Inviting you, come on!
Nourishing you with dreams.
Even when the rain is that gentle and quiet,
Dripping all along.

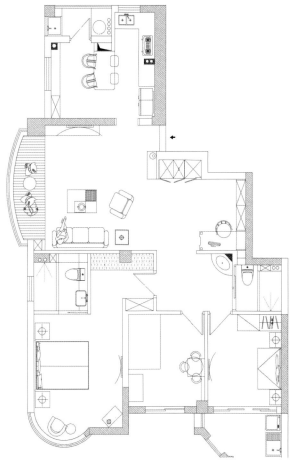

摩登样板间 III
新田园

玲珑
Exquisite Residence

设 计 师：严敏

Designer: Yan Min

美式田园风格除了给人以田园诗般的感受外，还有着高贵、奢华的空间印象。本案在美式风格的基调下，营造出田园般的诗意氛围。借着室内空间的解构与重组，缔造出一个令人神往的写意空间。

任何装修风格的前提都是要保障生活的舒适性，空间要体现出业主丰富的生活阅历，运用材质、色彩等来表达丰富的空间内涵。本案的设计彰显出空间的安逸感，又有着流行设计元素的搭配，巧妙的软装设计让人耳目一新。

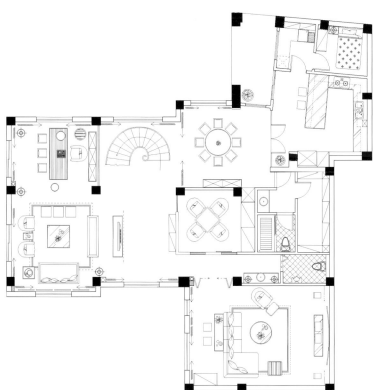
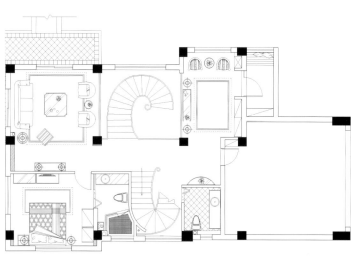

MODERN SHOW FLAT III NEW PASTORAL **189**

Apart from producing some pastoral poem like sensations, the American style pastoral style has some noble and luxurious space impressions. Under the tone of American style, the project produces some idyllic poetic atmosphere. With deconstruction and regrouping towards the interior space, the designer creates a fascinating enjoyable space.

The premise for any decorative style is to guarantee the comfort of life. The space shall present the abundant life experiences of property owner, and present abundant space connotations with materials and colors. The design of this project presents the space's easy and cozy feel, yet with collocations of trendy design elements. The ingenious soft decorative design produces a new appearance for the space.

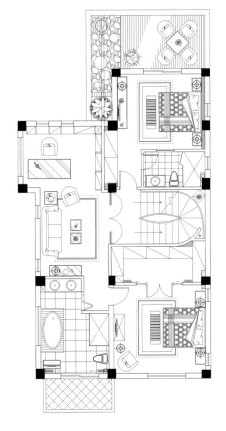
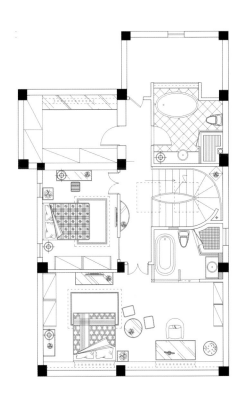

MODERN SHOW FLAT III NEW PASTORAL

MODERN SHOW FLAT III **NEW PASTORAL**

清浅时光·鸿雁名居
Light and Mild Time·Hongyan Mingju Residence

设计公司：北岩设计
设 计 师：于园
项目面积：90 ㎡
主要材料：硅藻泥、木板墙、复合地板等

Design Company: Beiyan Design
Designer: Yu Yuan
Project Area: 90 m²
Major Materials: Diatom Ooze, Wood Siding Wall, Composite Floor

考虑业主的预算和家庭生活的需求，设计师简化了硬装，让色彩大行其道。推开门，跳入眼帘的就是大红底色与绿色相互映衬的沙发，对比中的冲突反而使空间的个性得到释放。

为了节约餐厅的面积，设置了卡座，俏皮地配合两把红色的餐椅。铜质的灯具，让空间更加有质感。设计师寻觅很久的彩条窗帘，让空间的色彩活络起来。

主卧室中，设计师毫不犹豫地选择了立柱为红色的双人床，使红色成为一个流动的空间向导，花色复杂的窗帘使空间亮丽起来。设计师追随着空间的感觉，没有规则，没有必须，随心而为，寻找那份安静与安然。

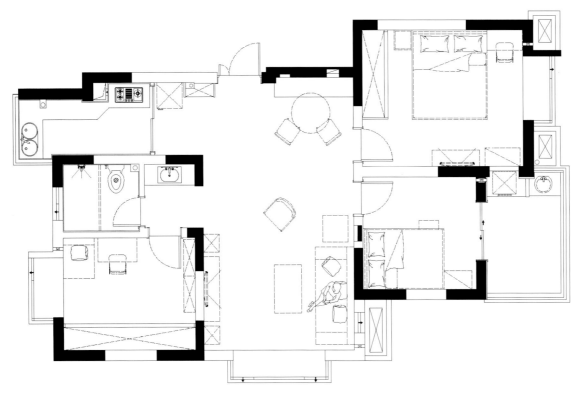

Considering the property owner's budget and requirements for family life, the designer simplifies the hard decorations and makes use of colors on a large scale. Opening the door, what comes into eyes first is the sofa of grand red color and green color echoing each other, with conflicts in comparison which release the features of the space.

In order to save the space of dining hall, the designer arranged a set seat here, in mischievous collocation with two red dining chairs. The copper lighting accessories make the space full of texture. The colorful curtains that the designer has been looking for in quite a long time make the space colors become dynamic.

Inside the master bedroom, the designer unhesitatingly selects double bed with red pillars, making red become the flowing space guidance and curtains of complex patterns light up the space. The designer follows the space feel, and searches for the tranquility and peace, with no rules, no necessities, just following the heart.

摩登样板间 III
新田园

世茂四期 5 幢
Shimao Property, Phase 4, Building 5

设计公司：大墅尚品·由伟壮设计	Design Company: Dashu Shangpin•Zhuang Design
设 计 师：由伟壮	Designer: You Weizhuang
软装设计：翁布里亚专业软装机构	Soft Decoration Design: Umbria Professional Soft Decoration Institution
软装设计师：王一飞	Soft Decoration Designer: Wang Yifei
项目地点：江苏常熟	Project Location: Changshu of Jiangsu Province
项目面积：88 ㎡	Project Area: 88 m²
主要材料：乳胶漆、饰面板擦色、欧松板勾缝、马赛克等	Major Materials: Emulsion Paint, Veneer Paint, OSB Board Jointing, Mosaic Tile

这是一套平层小公寓，乡村田园的风格使它显得温馨而惬意。在色调上，以暖色调为主，冷色调穿插其间，大面积的暖色调使生活在其中的人感到放松与舒适。对小户型面积的居室来说，居室功能空间的合理划分尤为重要，本着功能至上的原则，设计师将整体空间合理有度地规划开来，使之开合有度。

在灯饰的选择上，设计师选用了田园风格做旧的木质与铁艺相搭配的灯，顾及了田园风格的整体性，而且还与其他造型相互融合，使整体上充满自然的气息。

MODERN SHOW FLAT III NEW PASTORAL

This is a small flat apartment and the countryside pastoral style makes it appear warm and pleasing. The color tone focuses on warm color tone, with cold colors alternating inside. The large area warm color tone makes people living inside feel the relaxation and comfort. For residences of small house type, the appropriate segmentations of residential functional space is quite important. Based on the principle of functional supremacy, the designer properly plans the whole space, entrusting it with appropriate opening and closing design.

As for the selection of lighting accessories, the designer selects pastoral style archaized wood and matching iron art lights, taking into account the integrity of pastoral style, while integrating with other formats, making the whole space be full of natural atmosphere.

摩登样板间 III
新田园

乡居岁月
Time in the Village

设计公司：一空设计事务所
设 计 师：沈一
项目地点：福建福州
项目面积：107 ㎡
主要材料：有色乳胶漆、硅藻泥、橡木地板、仿古砖、进口壁纸

Design Company: YKON Design Studio
Designer: Shen Yi
Project Location: Fuzhou of Fujian Province
Project Area: 107 m^2
Major Materials: Colored Emulsion Paint, Diatom Ooze, Oak Wood Floor, Imported Brick, Imported Wallpaper

设计师希望业主在享受休闲居住空间的同时，也能感受到美式生活空间的文化内涵。因此，设计师用感性的语言诠释了美式乡村风格的概念，而且在空间感受的表达上有了进一步的升华。

在客厅的整体色调上，以暖黄色为基调，渲染出空间的大自然氛围。条纹相间的窗帘增加了空间的活跃感。餐厅、书房、卧室的设计也延续了客厅的基调，总体上给人以明媚的阳光感。设计师通过这种设计手法赋予这个空间一种全新的色彩体验，让生活不再单调。

The designer hopes the property owner to enjoy the cultural connotations of American style living space while enjoying the leisure residential space. Thus, the designer uses sentimental language to interpret the concepts of American style countryside style, while acquiring further sublimation in space sensations.

The whole color tone of the living room focuses on warm yellow color, setting off the natural atmosphere of the space. The curtain of stripes adds to the dynamic feel of the space. The design of living room, study and bedroom continues the tone of living room, while presenting to people some bright sunshine feel. With this kind of design approach, the designer entrusts the space with some brand new color experiences, making life not that monotonous.

MODERN SHOW FLAT III NEW PASTORAL

香江枫景 Fragrant River Scenery

设计公司：观云设计机构
设 计 师：林元君
项目地点：福建福州
项目面积：107 ㎡
主要材料：仿古砖、金刚板、松木板、乳胶漆、文化石、防腐木

Design Company: Guanyun Design
Designer: Lin Yuanjun
Project Location: Fuzhou of Fujian Province
Project Area: 107 m²
Major Materials: Archaized Brick, Laminate Flooring, Pine Board, Emulsion Paint, Cultural Stone, Antiseptic Wood

本案并没有赋予空间以特定的风格，根据空间本身的特点及业主的需求，设计显得随性而自然，在此可以找到田园风情，也可以找到欧式风情的影子。

空间选择实木地板，表现清新自然的空间感受。在空间色彩上，整体以淡雅的颜色为主，在局部区域搭配深色调，这样使得空间更加具有层次感。另外，部分功能区合理地利用了外部的自然光线，窗户的位置成为空间表现的重要保证。休闲阳台体现出浓郁的乡村格调，栅栏设计则表现出深厚的田园风情。

入户有一个露天场地，以防腐木地板铺设出一个休闲场地，既起着入户玄关的过渡作用，又营造出良好的生活情调。闲暇之时，一壶茶，一本书，就是一个世界。

This project does not entrust the space with peculiar style. The design appears natural and casual based on the space's own features and the needs of property owner. Here you can find the pastoral charms, as well as traits of European style.

The space selects solid wood floor to present fresh and natural space sensations. As for space colors, the whole space focuses on light and elegant colors, matched with dark color tones in some parts, which allow the space to have layer feel. Other than that, part of

the functional spaces properly makes use of exterior natural lights, while the location of window is an important guarantee for space expressions. The leisure balcony represents intensive pastoral style, and the fence design represents profound pastoral charms.

There is an open-air site at the entrance area, with a leisure area produced out of pavement of antiseptic wood floor board, which not only acts as transition space for the indoor hallway, but also produces great life charms. During the leisure time, with a pot of tea and a book in hand, you can find here a little world.

摩登样板间 III
新田园

太子山庄某住宅
One Residence of Taizi Villa

设计公司：深圳市伊派室内设计有限公司
设 计 师：段文娟
项目地点：广东深圳
项目面积：100 ㎡

Design Company: Shenzhen Yipai Interior Design Co.,Ltd.
Designer: Duan Wenjuan
Project Location: Shenzhen of Guangdong Province
Project Area: 100 m²

春暖花开，幸福小家

有人喜欢简洁明快的现代简约风，有人喜欢高贵典雅的美式风格，本案业主喜欢的是休闲而又不失个性的田园风格。设计师通过软装中的色彩与材质的搭配表现不同的空间气质与主题，为空间带来清新、自然的质感，使田园气息在现代的居室中散发出独特的空间表情。

本案户型结构功能明确、线条简洁，家具的选择简约而大气，干净清爽的气质给人们全新的视觉享受，有一种舒适、自然、休闲的味道。

Happy Family in Warm Family

Some like concise and brisk modern style, some like noble and elegant American style, while the property owner of this project likes leisurely yet peculiar pastoral style. Through colors and materials collocations in soft decoration, the designer represents different space temperaments and themes, creating some fresh and natural texture for the space, making pastoral atmosphere send out some peculiar space expressions in modern residence.

The house type structure has clear functions and concise lines. The selected furniture are concise and magnificent, which clear and fresh temperament leaves people with brand-new visual enjoyments, with some cozy, natural and leisurely tastes.

银城西堤国际某宅
One Residence of Yincheng Xidi International

摩登样板间 III 新田园

设计公司：北岩设计
设计师：于园
项目地点：江苏南京
项目面积：101 ㎡
主要材料：壁纸、罗马柱、实木地板、进口仿古砖等

Design Company: Beiyan Design
Designer: Yu Yuan
Project Location: Nanjing of Jiangsu Province
Project Area: 101 m²
Major Materials: Wallpaper, Roman Pillar, Solid Wood Floor, Imported Archaized Brick

家不再是苍白冰冷墙面与死气沉沉的地面的组合体，给家一种情调，给家一种精彩，在复古风里面寻找那些遗失的美好，发掘出别样的精彩是本案设计的宗旨。

在开放的空间里以暖色调为主，复古图案的米色壁纸，白棕木饰面墙裙，对应深色的复古浮雕面装点的电视墙，小型罗马柱隔断突出了设计的亮点。在餐厅部位将仿古瓷砖对贴，与厨房遥相呼应，餐厅古铜色的稻穗造型灯下，同色调的壁纸与木饰面相对比、相融合，凸显着业主不俗的品位。

Home is no longer a combination of pale and cold wall and lifeless ground, giving home some charms, and some splendidness. We are searching for the lost beauties in the antique style and it is the tenet of this project to find out some different splendidness.

The open space focuses on warm color tone, and the beige wallpaper of antique pattern, white sappan wood veneer wainscot echoes the TV background wall decorated with dark archaized relief surface, while the little Roman pillar partition highlights the design. The dining hall pasted the archaized ceramic tiles in some opposite way, echoing the distant kitchen. In the dining room, under the copper color light of rice ear format, the same color wallpaper is contrasting and integrating with the wood veneer, highlighting the uncommon tastes of the property owner.

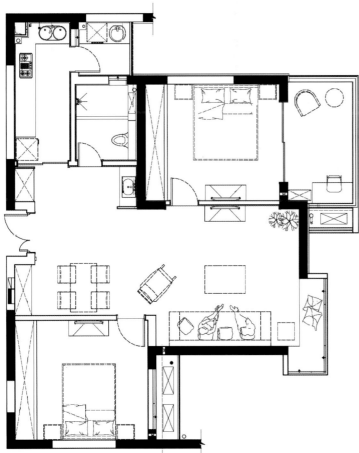

托乐嘉 Talege

设计公司：宇泽设计工作室	Design Company: Yuze Design Studio
设 计 师：肖为民	Designer: Xiao Weimin
项目地点：江苏南京	Project Location: Nanjing of Jiangsu Province
项目面积：150 ㎡	Project Area: 150 m²
主要材料：仿古地砖、马赛克、水曲柳饰面、麻质面料	Major Materials: Archaized Floor Tile, Mosaic Tile, Ash-tree Veneer, Linen Fabrics

设计师捕捉光线、追求色彩，为空间带来了疏朗开阔的视觉感受。黄与绿的组合配合家具的做旧处理表现出自然的美感；马赛克的镶嵌与彩砖的拼贴带来精致华丽的美妙质感；以点代面的装饰效果让空间散发出清悠的田园气息与尊贵的文化品位。

在空间的任何一个角落，都能感受到"开轩面场圃，把酒话桑麻"的意境，体会到主人悠然自得的生活和阳光般明媚的心情。

The designer captures light and colors to bring to the space expansive and bright visual sensations. The combination of yellow and green matches the archaized style furniture, creating some natural aesthetic feel for the space. The inlaid mosaic tiles and colored bricks parquet bring some delicate and magnificent texture. The decorative effects with spots replacing the surface make the space send out some leisurely pastoral atmosphere and noble cultural tastes.

No matter you are standing in what corner, you can always feel the artistic conceptions of "Opening the windowing, inviting the exterior views inside, enjoying the wine in hand, we can talk about the countryside sceneries on and on," and experience the property owner's leisurely life and sunny moods.

左右阳光 Left and Right Sunshine

设计公司：宇泽设计工作室
设 计 师：肖为民
项目地址：江苏南京
项目面积：92 ㎡
主要材料：仿古地砖、水曲柳饰面板、水曲柳实木面板、马赛克

本案风格轻松随意，干净明朗，貌似不经意的搭配却使空间舒展开来，一切浑然天成般的精彩夺目。在设计上突破了形式主义，注重功能性与实用性；在工艺上最大程度地保留了木质本身的自然色彩与天然纹理，没有张扬之感，展现出一种朴素、清新的原生态之美。

Project Location: Nanjing of Jiangsu Province
Designer: Xiao Weimin
Project Area: 92 m²
Major Materials: Archaized Floor Tile, Ash-tree Veneer, Ash-tree Solid Wood Panel, Mosaic Tile

The project style is relaxing and casual, clear and bright. The seemingly casual collocations spread inside the space. All is like nature itself, dazzling and brilliant. The design breaks the formal style, focusing on functionality and practicality. In technical sphere, the design maintains the wood texture's self natural colors and natural pattern to the largest extent, with no showy feel, and presenting some primitive and fresh original ecological beauty.

天正湖滨
Tianzheng Lakeshore

设计公司：振勇设计师事务所
设 计 师：冯振勇
项目地点：江苏南京
项目面积：170 ㎡

Design Company: Zhenyong Design Studio
Designer: Feng Zhenyong
Project Location: Nanjing of Jiangsu Province
Project Area: 170 m²

美式新古典风格的家具、仿古风格的地砖、弧形的墙面造型等都彰显出本案浓淡相宜的美式新田园风格的设计概念。淡黄的整体基调明快而雅致，使整个室内空间像是充满了春天的气息，局部点缀的宝蓝色更是起到画龙点睛的作用，就像是专为这春天而吹来的春风。

美式风格的家具一反古典风格的厚重、沉闷，体量不再一味求大求重，造型也不再繁复；材质不再以木质、皮质为主，而是多种材料并行，营造出一种轻灵、透亮的空间，既彰显浪漫的品质，又使空间充满清新的气息。

The American Neo-Classical style furniture, archaized style floor tiles, and arch wall formats all represent the American new pastoral style design concepts of appropriate lusters. The light yellow whole tone is brisk and elegant, filling the whole interior space with spring atmosphere, while the sapphire blue color in some parts makes the finishing point, just like spring breeze especially for the spring.

The furniture of American style removes the heavy and dull traits of classical style, the scale does not seek for grandness or heaviness blindly and the format is no longer complex. The materials are no longer focusing on wood and leather, yet with multiple materials, producing a bright and brisk space, which not only present some romantic quality, but also make the space be filled with fresh atmosphere.

摩登样板间 III
新田园

中冶虞山尚园
MCC Yushan Shangyuan Garden

设计单位：张之鸿设计事务所
设 计 师：张之鸿
项目面积：250 ㎡
主要材料：肌理涂料、仿古砖、美国红橡木

Design Company: Zhang Zhihong Design Studio
Designer: Zhang Zhihong
Project Area: 250 m²
Major Materials: Texture Coating, Archaized Brick, American Red Oak

本案在原来的布局上没有做过多的改动。因为是上、下两层的户型，我们就将厨房移到了地下，将原来厨房的位置改成了卫生间与洗衣间。原来相对较狭长的外卫做成了走道与储物空间。

我们要营造的是浓郁的美式乡村的感觉，所以采用深褐色为主色调；用木结构来模拟美式乡村房屋的建筑特色；用圆拱门和仿古砖及仿古地板这种材料来营造美式乡村的粗犷风格。在走廊中加入古典欧洲建筑元素，将走廊做成"十"字形，使居室具有复古的美式气质。

MODERN SHOW FLAT III NEW PASTORAL

This project does not make too many changes on the original layout. As it is a house type with two floors, we moved the kitchen to the lower floor and changed the original kitchen location into washroom and laundry room. The original comparatively long and narrow outside washroom is regenerated into a corridor and storage space.

What we want to create is the intensive American countryside feel, thus we apply dark brown color as the tone color and make use of wood structure to simulate the architectural features of American countryside houses. We

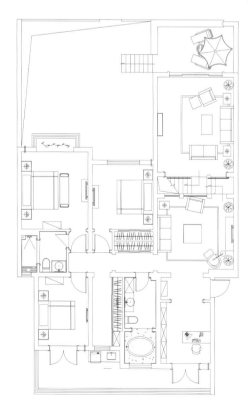
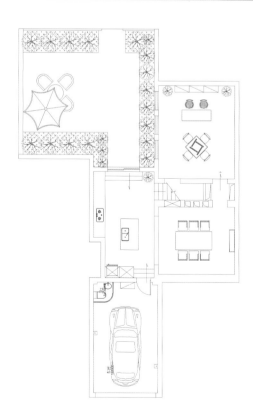

use materials such as round arch, archaized bricks and archaized floorboard to create the rugged style of American countryside. We incorporate classical European architectural elements into the corridor and make it present cruciform, making the residence acquire antique American temperament.

皇冠国际 Crown International

设计公司：方块空间
设 计 师：蔡进盛
项目地点：江西南昌
摄 影 师：邓金泉

Design Company: Fang Kuai Space
Designer: Cai Jinsheng
Project Location: Nanchang of Jiangxi Province
Photographer: Deng Jinquan

本案在色调上清新透亮，客厅中运用了白色墙板线框的造型，结合清新的浅蓝色、温暖的黄色作为点睛色，营造出温馨浪漫的风情空间，整体显得干净而优雅。

在居室的整体设计上，设计师没有采用传统的过于奢华的格调，而是加入一些新的创意与想法，将古典的元素融入到现代的风格中。装饰线条都极为简洁而没有丝毫繁复感。

设计师巧妙地把欧美的复杂线条进行抽象简化，运用一些新的装饰元素与摆饰，流露出对自然的向往，表达了一种恬淡、洒脱的生活方式。

This project is fresh and transparent in color tone. The living room applies the format of white wallboard wire-frame, combining fresh light blue and warm yellow colors as the highlighting elements, thus producing a warm and romantic charm space, with the whole space appearing clear and elegant.

As for the whole design of the residence, the designer does not apply the much too luxurious traditional tones, yet adding some novel innovations and ideas, and incorporating the classical elements into the modern style. The decoration lines are all quite concise, with no complex feel at all.

The designer ingeniously abstracts and simplifies the European and American style complicated lines, applying some new decorative elements and ornaments, revealing people's longings for grand nature and displaying some free and easy lifestyle indifferent to fame or fortune.

中山星汇云锦3栋 03户型

Zhongshan Starry Winking, Building No. 3, House Type 03

设计公司：广州市韦格斯杨设计有限公司
项目地点：广东中山
主要材料：米黄石、仿古砖、马赛克、壁纸、白色木饰面

Design Company: Guangzhou Grand Ghost Canyon Designers Associates Ltd.
Project Location: Zhongshan of Guangdong Province
Major Materials: Beige Stone, Archaized Brick, Mosaic Tile, Wallpaper, White Wood Veneer

田园风光一直就是浪漫、温馨的象征，是浪漫爱情与大自然结合的体现。本案采用的是浪漫田园式的设计风格，着意营造出一种休闲、浪漫、阳光的生活氛围。

在空间布局上，把主卧与客卧做了连通式的处理，原先狭小的主卧变得宽敞了，同时拥有了一个独立的衣帽间与主卫，意在打造一个浪漫、舒适的二人世界。

配饰上，茂盛的花艺、典雅的铁艺挂饰，充分衬托出空间的浪漫情调。加上空间丰富的色调，轻松休闲的氛围，使人流连忘返。

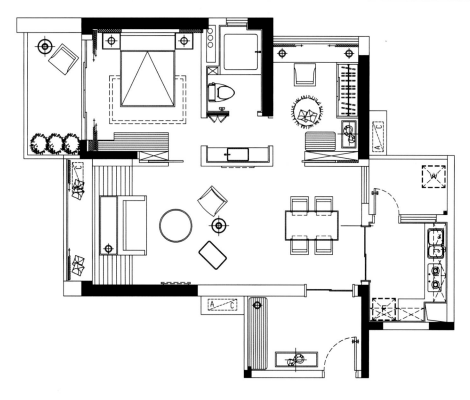

Idyllic scenery is the symbol of romance and warmth, as the representation of combination of romantic love and grand nature. This project applies romantic pastoral design style, creating some leisurely, romantic and sunny life atmosphere.

In space layout, the master bedroom and guest bedroom are connected with each other. The original narrow master bedroom is made broader, while with an independent cloakroom and a master's washroom. The designer intends to produce a romantic and cozy world for the couple.

As for ornaments, the lush floriculture, and the elegant iron hanging decorations fully set off the space's romantic tones. Accompanied with the space's rich color tones, the relaxing and leisurely atmosphere makes people hesitate to leave.

佳兆业鞍山水岸华府 GC-A3 样板间

Kaisa Anshan Waterfront Mansion, GC-A3 Show Flat

设计公司：广州市韦格斯杨设计有限公司

项目面积：74.8 ㎡

主要材料：爵士白大理石、黑白根大理石、瑞丽米黄大理石、木饰面、壁纸等

Design Company: Guangzhou Grand Ghost Canyon Designers Associates Ltd.

Project Area: 74.8 m²

Major Materials: Jazz White Marble, Black Marquina, Beige Marble, Wood Veneer, Wallpaper

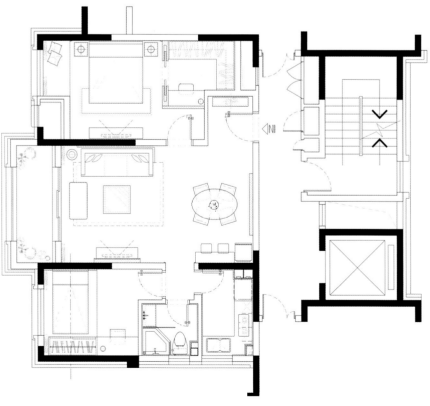

本案通过对"英式田园风格"的重新演绎，营造出悠闲、舒适、地道的轻调英式田园空间。尤其是壁纸的选用，客厅中淡绿色碎花压纹的壁纸，体现出家庭生活的轻松感。

主人房墙面以深红色的壁纸为主，儿童房以表现天真烂漫的童年生活为主题，还选用夜光壁纸。三个空间，通过同一材料，但是不同颜色的运用，使其整体统一，而又各具特点，创造出了宜居、舒适的家庭氛围。

Through reinterpretation of "English Pastoral Style," this project creates some leisurely, cozy and genuine English pastoral space. Especially the selection of wallpaper, the living room applies the wallpaper with light green floral pattern, representing the relaxation feel of family life.

The master bedroom's wall focuses on dark red wallpaper, while the children's room has the theme of innocent and wonderful childhood life, and with noctilucence wallpaper. Through the use of same materials of different colors, the three spaces attain integrity, yet with peculiar features respectively, creating livable and cozy family atmosphere.

中信红树湾
CITIC Mangrove Bay

设计公司：深圳三米家居设计有限公司
设 计 师：3米
项目地点：广东深圳
项目面积：154 ㎡

Design Company: Shenzhen 3mi Home Furnishing Design Co., Ltd.
Designer: 3mi
Project Location: Shenzhen of Guangdong Province
Project Area: 154 m²

我们从自然中来，天地是我们的父母。花儿、小草、树木、鸟儿犹如我们的兄弟姐妹，共同在大地母亲的怀抱中快乐成长。

在本案中，设计师将这些自然元素融入于设计中，打造出使人倍感亲切的生活环境，让人忘却烦恼，只为享受生活而存在。

本案以现代田园为主调，不同于传统田园风格的怀旧感，现代田园是在田园氛围里融入现代简约元素，营造一种更轻松、质朴、清爽的感觉。灰蓝的色调、花卉图案、木制家具、陶瓷工艺、绿色植物、简约的造型……在流畅的线条中流露出清爽的味道，成为都市中的"雅居"。

We come from nature, and the heaven and earth are our parents. Flowers, grass, trees and birds are like our brothers and sisters, and we are growing up with them in the arms of our parents happily. In this case, designers merge natural elements into the house and create an amiable living environment, which let people forget the troubles in daily life, but completely enjoy the joy of life.

This case focuses on modern garden style, but it is different from traditional rural style, which is nostalgic. The modern garden style blends modern concise elements into the rural atmosphere and presents an easy, plain and refreshing expression. The gray blue tone, floral patterns, wooden furniture, pottery, green plants and concise modeling, etc., all those elements form smooth lines and create a salubrious feel and an "elegant home" in the metropolis as well.

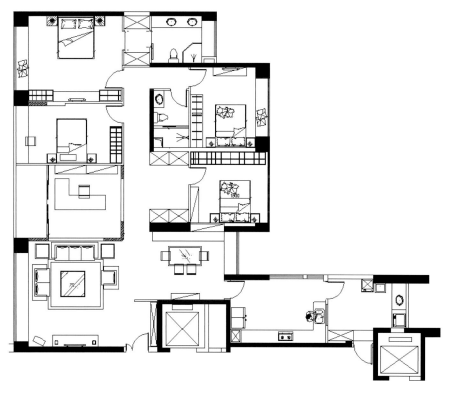

星梦奇缘
Star in My Heart

设计公司：上海映象
项目面积：350 ㎡

Design Company: Shanghai Impression Design
Project Area: 350 m²

星梦奇缘的房子是一栋可以容纳全家老小的住宅，位于上海市闵行区。这栋别墅的独特之处在于它有地下阳光室，面积与一楼相同。住宅设计巧妙结合了两种相应的建筑形式。整栋房子光线与色彩的控制是重中之重。夜晚是室内灯光展现魅力的时刻，宽敞开阔的起居空间清爽怡人。精心挑选的家具与面料，配合怡人的灯光，营造出了宁静慵懒的室内氛围。

不规则的空间可以产生多变的趣味性，为了保持这一优势，采用了开放式的交流结构，在地面上用地砖做区域上的划分，这样就有了静态空间与动线路径上的分别，是隐喻而有效的空间划分手法。

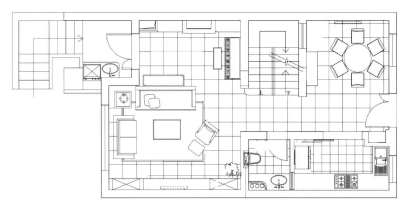
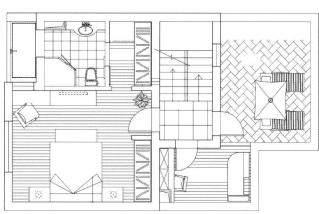
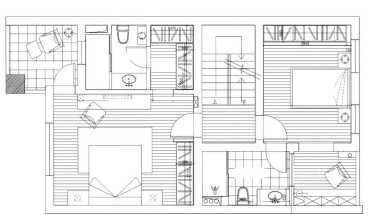
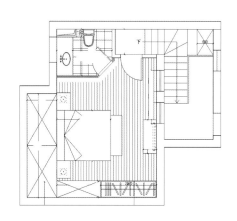

MODERN SHOW FLAT III NEW PASTORAL **283**

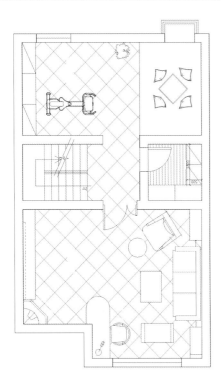

This project is located in Shanghai's Minhang District, which can accommodate the whole family, including young and old. The peculiar part of this villa is that it has an underground sunshine room, with the same area as the first floor. The residential design ingeniously combines these two corresponding architectural formats. The control on the whole residence's lights and colors is the priority among priorities. Night is a time when the interior lights display their charms. The spacious and broad living room is so pleasing and refreshing. The painstakingly selected furniture and materials match the pleasing lights, creating some tranquil and lazy interior atmosphere.

The irregular space could produce varying interests. In order to maintain that advantage, the designer applies open style communication structure, having regional segmentation on the ground with floor tiles, which forms separation between tranquil space and moving lines, as an implicit yet efficient space division approaches.

摩登样板间 III
新田园

蓝湖郡联排别墅
Blue Lake County Townhouse

设计公司：重庆十二分装饰工程设计有限公司
设 计 师：田艾灵（田芬）
项目面积：251 ㎡
主要材料：壁纸、复古地砖、实木地板、彩绘玻璃、手工制作雕刻家具、天然石材

Design Company: Chongqing Shi'er Fen Decorative Engineering and Design Co., Ltd.
Designer: Tian Ailing
Project Area: 251 m²
Major Materials: Wallpaper, Archaized Floor Tile, Solid Wood Floorboard, Stained Glass, Hand-Made Carved Sculpture, Natural Stone

家如其人，找其味、定其调，因为房主夫妇都属于温润谦和的类型，设计师从其个性及谈吐中提炼出温婉、内秀的主题来展现居住空间的"润"。浅浅的蓝、淡淡的色、莹莹的瓷勾勒出这一方天地，如同一曲清浅天籁，悄然奏响，动人心弦。

设计师采用整体融合、一气呵成的自然写意手法，营造出惬意、舒适的空间氛围。同时，采用自然的浅灰蓝加米色色系，辅以纯净、明亮的窗饰，让室外的光线和风景更好地为室内环境衬底，使室内、外景色完美融合的同时，也更显亲近、自然。

MODERN SHOW FLAT III NEW PASTORAL

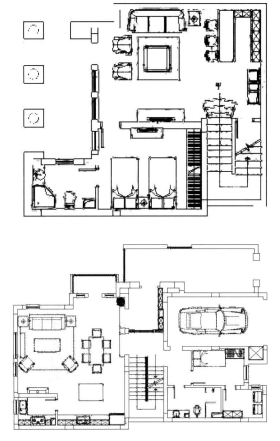

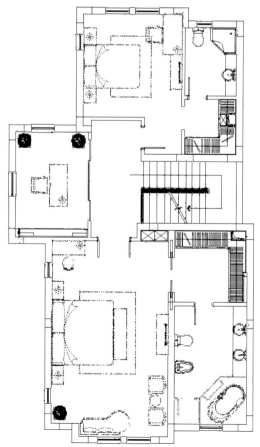

Home is like the person, and we are searching for its taste and tones. The couple are both mild and modest. From their characteristics and style of conversation, the designer extracts the gentle and restrained theme to present the "smoothness" of the residential space. Light blue, slight colors and dazzling china delineate this little world, which is like a musical chapter, deeply moving people.

The designer makes use of integral and coherent natural freehand approaches, creating the cozy and relaxing space atmosphere. While at the same time, the designer makes use of natural light gray blue and beige colors, accompanied with pure and bright window accessories, making the exterior lights and landscapes better set off the interior environment. While perfectly integrating interior and outside sceneries, the space appears more natural and intimate.

葱荣岁月·雅致如歌
Memorable Years like a Song

设计公司：北岩设计
设计师：于园
项目地点：江苏南京
项目面积：140 ㎡
主要材料：艾芙迪家具、进口壁纸、进口面砖等

Design Company: Beiyan Design
Designer: Yu Yuan
Project Location: Nanjing of Jiangsu Province
Project Area: 140 m²
Major Materials: FD Furniture, Imported Wallpaper, Imported Bricks

有人说:"家不是一个空间,而是一段时光"。设计师们雕刻的是一个优雅的容器,装载着一路走过的欢声笑语,记载一段如歌如诗的岁月。如果说家是心中永恒的港湾,那么它更是一道风景,在这里享受美食、分享快乐和劳动的乐趣……

在本居室中,设计师以清新亮丽的元素为整个家奠定了一个主色调,以完美的弧线、优雅的花型和精致的细节为家略施粉黛,最后配上精致、典雅的软装饰品,共同演绎一场视觉盛宴,令人眼前一亮。别样雅致的生活从此开始。

Some say that "Home is not a space, but a time." What the designer sculpted is an elegant container, loading the happy time all along the way and recording the time which is like a song, like a poem. If we say that home is an eternal harbor in heart, it is a landscape, where one can enjoy the gourmet, share happiness and enjoy funs of labor...

Inside the residence, the designer applies fresh and bright elements to set the tone color for the whole house, decorating the home with perfect arches, elegant floral pattern and delicate details, finally accompanied with delicate and elegant soft decorative objects, together creating a dazzling visual banquet. Some spectacular life starts here.

太古城 D 座 A 户型
Taigu City, Building D, House Type A

设计公司：马思威设计工作室	Design Company: Ma Siwei Design Studio
设 计 师：马思威	Designer: Ma Siwei
项目面积：120 ㎡	Project Area: 120 m²
主要材料：大理石、仿古砖、实木线条、墙漆	Major Materials: Marble, Archaized Brick, Solid Wood Lining, Wall Paint

时尚的线条与灵动的空间流线,是本案的灵魂。玄关没有做过多的修饰,蜂窝结构的酒柜体现了主人对美的感受。这种愉悦感使我们感受到了空间特有柔和气息。

在大体确定的空间格调的基础上,设计师把色彩的运用作为本设计的另外一个亮点。淡绿色的基调与仿古砖、灯饰,米黄色的沙发相结合,让我们感受到了生活的美好。

The soul of the project is fashionable lines and dynamic space streamlines. The hallway does not make too many decoration. The wine cabinet of honeycomb structure represents the master's sensations towards beauty. The pleasing feel makes us feel the soft atmosphere exclusive to the space.

Based on the approximately set space tones, the designer makes use of color as another highlight of the project design. The light green color tone is combined with archaized bricks, lighting accessories and beige sofa, which make us feel the beauty of life.

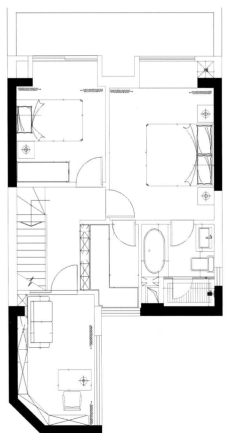
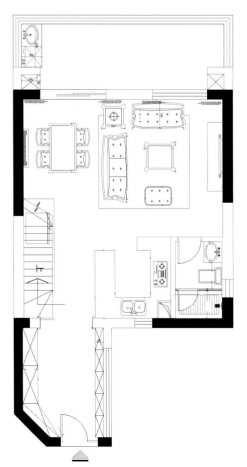

摩登样板间 III 新田园

颐慧佳园样板间
Yihui Jiayuan Show Flat

设计公司：圳銮想向（北京）装饰艺术设计有限公司
设计师：向东姝
项目地点：北京
项目面积：160 ㎡
主要材料：进口壁纸、仿古地面砖、涂料

Design Company: Beijing Zhenluan Xiangxiang Decorative Art Design Co., Ltd.
Designer: Xiang Dongshu
Project Location: Beijing
Project Area: 160 m²
Major Materials: Imported Wallpaper, Archaized Floor Tile, Coating

本案为美式新乡村风格的家居设计，不同于传统的美式乡村风格。新乡村风格打破了传统美式风格全部采用实木质感的家具及刻意做旧的岁月痕迹，穿插进各种花卉绿植、具有异域风情的饰品、极具田园气息的壁纸、带有自然韵味的家具及精美的铁艺制品，让它们在家居环境中充分地展现各自的魅力，营造出了一个舒适、浪漫、温暖的家庭氛围。

走进客厅就能感受到清爽明快的气氛，灰蓝色的壁纸上印有简洁的花纹，与壁纸同色系的沙发用格子图案点缀，温馨中流露出几分浪漫情调。绿色的窗纱与室内的绿色植物相映成趣，又增添了几许生机与活力。

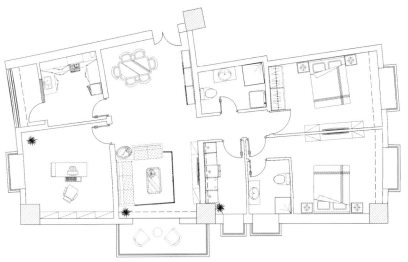

This project is for furnishing design of American new countryside style, which is different from traditional American countryside style. The new countryside style breaks the approaches of traditional American style wholly using solid wood texture furniture and antique-feel traits of time, interspersed with various kinds of flowers and plants, ornaments of exotic charms, wallpaper of extreme pastoral atmosphere, furniture of natural charms and exquisite iron

accessories, which fully display their respective charms in the residential environment and form this cozy, romantic and warm family atmosphere.

Upon stepping inside the living room, you can find the refreshing and brisk atmosphere. The grayish blue wallpaper is printed with concise pattern, and the same color sofa is ornamented with check pattern, revealing some romantic charms in the warm atmosphere. The green window curtain and interior green plants reflect the best in each other, adding some vigor and vitality.

MODERN SHOW FLAT III NEW PASTORAL 309

花里林居
Neighbor in the Flower

设计公司：深圳装饰工程有限公司
设 计 师：郑福明
项目面积：110 ㎡

Design Company: Shenzhen Decoration and Engineering Co., Ltd.
Designer: Zheng Fuming
Project Area: 110 m²

营造一个心灵的港湾，不需要多么华丽，也不需要多么绚烂。只是需要一个在你身心疲惫的时候可以依偎、高兴的时候可以手舞足蹈的地方。

在本案不大的空间中，设计师在简约风格中融入了一些装饰主义的特征，这样就给过于坚硬的外表增添了些许温情的氛围。简约的家具搭配饶有趣味的配饰，再加上稳重又不失活泼的色彩，使得这个居室充满着浓浓的人情味，使之真正成为一处灵魂的归属地。

在灯光设计上，光线让视线得到了最大化的延伸，丰富了空间的层次。在功能设计上勾勒出了一个简单而颇具个性的现代化生活空间，是一个满含设计师热情与智慧的设计作品。

A harbor for the hearts does not need to be gorgeous, nor splendid. You only need such a place where you can take a rest when you are tired, and you can dance happily when you feel great.

This is a place belonging to the psyche. Within the not too big space, the designer incorporates some decoration style characteristics in the concise style, which adds some warm senses to the much too hard appearance. Concise furniture is accompanied with fun ornaments, together with the sedate but lively colors, the whole residence is full of intensive human kindness, making the space become a destination for the psyche.

As for lighting design, the lights extend the visions to the utmost, enriching the space layers. The functional design creates a simple but peculiar modern life space, which is a design work full of the designer's passion and wisdom.

大华南湖公园世家

Dahua Group South Lake, La Park

设计师：孔鑫
项目地点：湖北武汉
项目面积：60 ㎡
主要材料：乳胶漆、壁纸、木地板等

Designer: Kong Xin
Project Location: Wuhan of Hubei Province
Project Area: 60 m²
Major Materials: Emulsion Paint, Wallpaper, Wood Floor

本案为60㎡的小户型，比较适合经济能力一般的城市白领。通过与业主的充分沟通，想要有较大的空间视野、储藏功能要强。

户型改造后，改变了厨、卫、卧室的动线，使空间分区更合理，充分利用过道资源。吧台起到承上启下的作用，既能满足与厨房的功能交互，又能作为多功能榻榻米区与卧室、客厅之间的过渡区，一举多得。榻榻米区能满足棋牌、茶室、睡眠区的功能需求。

沙发与床之间加上屏风隔断，既能当作沙发的背景墙，又能作为两个区域的隔断。在空间的储物功能上，充分利用每一个角落，如飘窗矮柜、吧台踏步旁的边柜、阳台储藏柜、榻榻米等，且风格统一不突兀。

This is a little house type of 60 m², quite fit for urban white-collars of average economic capabilities. Through sufficient negotiations with the property owner, we get to know that he wants comparatively large space views and powerful storage functions.

After house type regenerations, the moving lines of kitchen, wash room and bedroom are changed, making the space division become more appropriate, while making full use of the corridor space. The bar counter has the function of being a connecting link, which can not only meet with the functional interactions with the kitchen, but also act as the transitional area between multi-functional tatami and bedroom, living room, answering multiple purposes. The tatami area can meet with the functional requirements of chess and card area, tea space and sleeping area.

There is a screen partition between the sofa and bed, which not only has the role of sofa's background wall, but also is the partition between the two areas. As for space storage functions, the designer makes full use of each corner, such as bay window low cabinet, side cabinet along the bar counter's stepping, balcony storage cabinet and tatami, while with unified style.

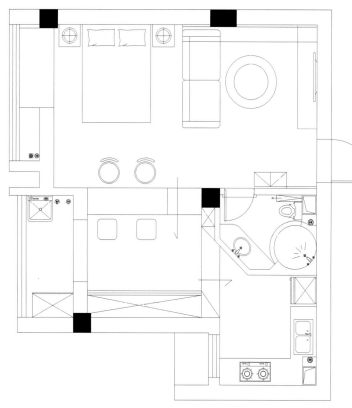

MODERN SHOW FLAT III NEW PASTORAL

本书在编写过程中，得到各位参编老师的倾力协助，特表示感谢，以下为参编人员名单（排名不分先后）：

庞一飞　殷正毅　代曼淇　李　扬　祝竞如　江　嘎　叶子丰　林元君　韩　松　陈　琼　金啸文
朱林海　蔡进盛　刘海涛　林函丹　张之鸿　孙　滔　杨旭光　马燕艳　由伟壮　黄　莉　任朝峰
刘　鹏　施海杰　蔡其辉　张喜波　于　园　李　健　颜　旭　段文娟　陈　熠　沈　一　林函丹
严　敏　肖为民　冯振勇　张之鸿　蔡进盛　田艾灵　马思威　向东姝　郑福明　孔　鑫

图书在版编目(CIP)数据

摩登样板间. 第3辑. 新田园 / ID Book 图书工作室编 —武汉：华中科技大学出版社，2014.9
ISBN 978-7-5680-0066-6

Ⅰ．①摩… Ⅱ．①I… Ⅲ．①住宅－室内装饰设计－图集 Ⅳ．①TU241-64

中国版本图书馆CIP数据核字(2014)第100151号

摩登样板间Ⅲ 新田园　　　　　　　　　　　　　　　　　　　　　　　　　ID Book 图书工作室 编

出版发行：华中科技大学出版社（中国·武汉）
地　　址：武汉市武昌珞喻路1037号（邮编：430074）
出 版 人：阮海洪

责任编辑：赵爱华　　　　　　　　　　　　　　　　　　　　　　　　　　　　责任监印：秦　英
责任校对：曾　晟　　　　　　　　　　　　　　　　　　　　　　　　　　　　装帧设计：张　艳

印　　刷：北京利丰雅高长城印刷有限公司
开　　本：965 mm×1270 mm　1/16
印　　张：20
字　　数：256千字
版　　次：2014年9月第1版第1次印刷
定　　价：338.00元　（USD 69.99）

投稿热线：(010)64155588-8000
本书若有印装质量问题，请向出版社营销中心调换
全国免费服务热线：400-6679-118 竭诚为您服务
版权所有　侵权必究